TABLE OF CONTENTS

CHAPTER 1 - INTRODUCTION

Learning Opportunities:

- *Deploying retail technology solutions no longer need to be dreaded.*

- *There is a proven method that will not only save time and money but will position anyone using this method as a deployment leader.*

- *Creating a Deployment Design is key to the success of any retail technology deployment.*

Everyone has been there: spending lots of money on acquiring hardware and software, hundreds of hours testing the technology solution, ensuring everything will work exactly as desired, when needed, to output the desired results. All so revenue can be increased, new customers can be acquired, and the brand's reach can be expanded.

Once the solution is ready, it's time to deploy it to all the stores. A vendor is selected that is believed to have the best track record with the specific type of deployment, contracts are executed, a project plan is developed, resources are all lined up, and, eventually, the green flag is dropped. Then, everything goes wrong, and within just one week the wheels fall off—missing equipment, no-show Deployment Technicians, stores are not ready, and many complaints from both store personnel and customers. All that hard work has gone down the drain.

Not only was the deployment a failure, but customers and Store Managers are upset. Additional costs for revisits, expedited shipping costs, overtime, wait time, out-of-scope work, and storage fees were incurred—all because the time was not taken to design the deployment.

Unfortunately, this happens so many times. Too much time and money are spent on the solution, but virtually neither is spent on designing the deployment itself. While this occurs in virtually all types of deployments, it is most devastating in a retail technology deployment. No matter how great and wonderful the technology, the project will be viewed as a failure if technology cannot be quickly deployed the first time with minimal or no interruption to business. Designing a retail technology deployment will not be easy, but it will set the project up for success if this process is completed before anyone ever steps foot in the store.

No one would ever sit down with their spouse to plan the building of their dream home without bringing in an expert to help plan, draw up a blueprint, and go through all the details of the construction. They would not only pick the number of bedrooms and bathrooms, but would also select the paint color, carpet design, and even the kinds of knobs to put on the kitchen cabinet doors. So why would anyone skip this step with deploying a technology solution? All those hours spent solutioning and testing the technology, why shouldn't the same be done with the deployment?

While all the intricacies of the technology and the software may be known, it may be uncomfortable when considering all the steps required when designing a deployment. It feels like there is so much that is unknown. There are so many questions: What if something is missed? What about areas where there is no expertise? What all should be included or not included?

If these scenarios and questions mentioned throughout this chapter resonate, then this book has the answers. It covers everything to consider when designing a retail technology deployment. While it may seem overwhelming initially, if bite-sized pieces of the information given are taken, chewed thoroughly and thoughtfully, there is no doubt success will be achieved.

Key Takeaways:

- *Too many deployments experience significant issues because they were only planned—not designed.*

- *There is a proven way to create Deployment Designs.*

- *Set the project up for success by creating a Deployment Design regardless of the size or scope.*

CHAPTER 2 – DESIGN VS. PLAN

Learning Opportunities:

- The difference between a Deployment Plan and a Deployment Design.

- The importance of creating a Deployment Design regardless of the size or complexity of the deployment.

- The cost benefits that can be realized from developing a Deployment Design.

Merriam-Webster's Dictionary defines *plan* as "to arrange a method or scheme beforehand for any work, enterprise, or proceeding" and defines *design* as "to prepare the preliminary sketch or the plans for (a work to be executed), especially to plan the form and structure of".

Wikipedia elaborates on the definition of *design* as "Designing often necessitates considering the aesthetic, functional, economic, and sociopolitical dimensions of both the design object and design process. It may involve considerable research, thought, modeling, interactive adjustment, and re-design".

While the differences in definition may appear subtle, the reality for Retailers is that simply planning a technology deployment is not enough. A technology deployment must be designed. A Deployment Plan outlines what is going to be done. Whereas, a Deployment Design specifically details how it will be done. Here are some examples.

Imagine a project to deploy updated wireless infrastructure to 800 of a 1,500-store portfolio, and the project must be completed in six months. Although this seems straightforward enough, consider the following illustrations of how this would look in a Deployment *Plan* versus a Deployment *Design*.

Within a Deployment *Plan* the time frame during which the deployment must be completed is established. Within a Deployment *Design* the business drivers for the requested time period and the financial impact to the business if missed are documented. The actual schedule of which stores will be deployed when is also designed, with considerations for preferred in-store times, blackout periods, ramp periods, and availability of store personnel. Finally, an associated milestone schedule and process for schedule changes is included.

Within a Deployment *Plan* a list of equipment to be deployed is obtained. Within a Deployment *Design* all the specific details of the equipment to be deployed are obtained, including make, model, size of device including the box, and weight. Associated specifics include how and who will procure and ship the equipment, how Dead-On-Arrival (DOA) equipment will be handled, and how legacy equipment will be dispositioned.

Within a Deployment *Plan* the task of completing a site survey for each store is included. Within a Deployment *Design* a schedule of site surveys is designed by store with assigned Deployment Technicians for each, communications to be used (e.g. call ahead script, email, intranet post, etc.) by Store Operations, checklist or survey script to be used by the Deployment Technicians, length of time to complete the survey, method of collecting information and transmitting back to the Project Team, and how results will be stored and used in the deployment process.

Within a Deployment *Plan* a task is included to develop a deployment script. Within a Deployment *Design* fully documented step-by-step instructions for the Deployment Technician to follow during the deployment is developed. This script includes all the details and is 100% accurate. This requires iterative testing until every step can be executed flawlessly, including every single aspect of the deployment from the time the Deployment Technician enters the store until the time he or she leaves.

From these few examples, it is clear that a Deployment Design must include all the details from the beginning to the end of the deployment. No detail should be considered too small. Granted, some minute detail may be missed, or some processes may have to be tweaked during the actual deployment, but with such a detailed design in place adjustments will be able to be made quickly, causing little or no negative impact to the deployment.

The biggest question is how will designing deployments positively impact the bottom line? The biggest impact of designing the deployment is that it eliminates the unknowns and provides a mechanism to conduct dress rehearsals for the deployment and all related processes and procedures. Specifically, *Figure 2-1* outlines just a few of the benefits that can be achieved when a Deployment Design is used.

Example Benefit	How Benefit Achieved
Reduction or elimination of reschedules and revisits	By designing Readiness Assessments throughout the deployment, the status of the store is known ahead of time
Maximized deployment time in the store	Having a step-by-step, fully tested Deployment Script eliminates Deployment Technicians "figuring it out" or interpreting instructions incorrectly while in the store
All deployment participants know what is expected which eliminates rework or the negative impact of miscommunications	With all processes and procedures fully documented, expectations or properly set and all participants know how to handle exceptions expeditiously and efficiently

Example Benefit	How Benefit Achieved
Minimized impact to the store and increased customer satisfaction for the end users	With a fully detailed Deployment Script, Deployment Technicians get in and get out much faster and with it being correctly done the first time, there is greater acceptance and satisfaction
Identify gaps and omissions early and with enough time to correct	A Deployment Design requires the review of all details related to the deployment which means all the potential issues that were not thought of during the technology solutioning phase will be quickly discovered
Eliminate costs associated with expediting resources and equipment	By designing the appropriately placed checkpoints will allow proactiveness and identification of issues while there is time to take corrective action without additional costs

Figure 2-1

While the benefits of a Deployment Design can be many depending on the different types and complexities of deployments, it might be helpful to outline some common use cases that show the direct savings that can be gained.

Here's an example. Rose's Reads is a national Retailer based in the Midwest that has a portfolio of 1,680 book stores. Established in 1950 as a very small local bookstore, they now offer a wide array of amenities to their customers including a wine and craft beer bar offering monthly tasting parties, a café with snacks and an extensive menu of coffee selections, meeting rooms for book clubs, book signings and corporate events, quiet lounge areas where customers can curl up with a good book or work on their mobile device, a children's area that provides short-term sitting services, and a sound-proof room where customers can listen to the newest music of any genre. They not only offer the traditional checkout model, but they also provide self-checkout and buy online pickup in store services. Rose's Reads is committed to creating an environment that is inviting and gives their customer's reasons to stay for an extended period and return as a place where they feel a sense of community.

Obviously, Rose's Reads has a great deal of technology needs and requires ongoing maintenance as well as break fix services when issues arise. With the nature of the business they are sensitive to any interruption to their store, so they tend to schedule this type of work before or after hours. One recent project they completed was to upgrade their wireless network. This included replacing all the wireless access points which required a lift to install, upgrading the store's switch, and replacing the wireless controller. To obtain better coverage with the new system, they added wireless access points and moved the position of existing locations, which required additional cabling.

They decided they would procure, configure, and ship all the equipment to the stores where they would store the equipment in the back office of each store. Then they would contract with a third-party provider to send a Deployment Technician to each store based on a schedule Rose's Reads provided them to complete the installation, turn up, and test out.

The project was completed, but experienced significant cost overruns and were two months past the desired completion date. They decided to complete a review into how the project cost so much and went so long. A root cause analysis revealed they spent over **$244,000** that could have been avoided if they had completed a Deployment Design. *Figure 2-2* outlines the results of this analysis.

Issue:	Deployment Technicians had to spend additional time looking for misplaced or missing equipment.
Root Cause:	Store personnel were not provided with instructions on where to properly store and secure the equipment until the Technician arrived.
Results:	Produced out-of-scope charges for 170 hours of Deployment Technician time to either search for equipment or wait for store personnel to locate the equipment.
Impact:	170 hours * $75/hour = **$12,750**

Issue: Some equipment was misconfigured and would not work properly once installed in the store.

Root Cause: The configuration for the equipment is store-specific. However, there was not a process in place to label the devices in a manner that everyone would know what store the equipment was to be shipped to. Equipment was inadvertently mixed up and shipped to the wrong stores.

Results: Deployment Technicians had to reconfigure devices in the field which produced OOS charges for 340 hours.

Impact: 340 hours * $75/hour = **$25,500**

Issue: Cabling required for additional and relocated wireless access points was not completed prior to the scheduled deployment date.

Root Cause: While Rose's Reads provided the schedule to the third-party provider, they did not implement a change management process to properly change and communicate schedule changes. Additionally, call aheads were not performed to ensure that the cabling work was completed prior to the arrival of the Technician.

Results: A revisit charge was incurred at $150 for each occurrence and the store had to be rescheduled later in the master schedule.

Impact: 118 revisits * $150 = $17,700

Added 10 days for rescheduled stores and $20,000 additional for resources to remain after the original scheduled completion date of the project.

Total Impact: **$37,700**

Issue:	The lifts were rented from a third-party rental company and they were not picked up in a timely fashion once the deployment was complete.
Root Cause:	There was no process in place to ensure that the rental company was contacted upon completion and no follow up to ensure the lifts had been picked up as scheduled.
Results:	504 extra days of lift rentals.
Impact:	504 days * $200/day = **$100,800**

Issue:	After 500 stores the deployment was consistently not being completed prior to store opening so it was decided to start the deployment one hour earlier.
Root Cause:	A detailed step-by-step deployment script was not thoroughly tested and optimized.
Results:	This required the Store Manager to start an hour earlier, which meant they would have to be paid overtime.
Impact:	1,180 stores * ($38[a] * 1.5) = **$67,260**
	[a] = Average Store Manager Hourly Rate

Total Impact:	**$244,010**

Figure 2-2

While everyone may plan for a deployment, few take the time to script a formalized design. The Deployment Design should be part of the Deployment Plan, along with the proper amount of time required to complete. Depending on the size and breadth of the project, this could be a significant amount of time, but the time and money saved makes the effort worth it. In today's highly competitive retail environment it is critical to deploy technology flawlessly, expeditiously, and seamlessly while ensuring that the store continues operationally. Developing a Deployment Design will ensure all these objectives are met.

Key Takeaways:

- *A Deployment Plan communicates what work will be done. Whereas a Deployment Design communicates specifically how the work will be done.*

- *A Deployment Design must include all the step-by-step details prior to any deployments being completed in the store.*

- *No Retailer can afford NOT to complete a Deployment Design.*

CHAPTER 3 – DESIGN WORKSHOP

Learning Opportunities:

- *The importance of conducting a Design Workshop.*

- *How to conduct a Design Workshop with the Project Team.*

- *How the output of the Design Workshop is used in the Deployment Design.*

A Design Workshop is a great way to facilitate and organize the development of the Deployment Design with the core Project Team. Either, an experienced Deployment Designer can be brought in to facilitate the discussion and fully document the design or the members of the Project Team can be used to complete the design.

While less complex deployments can be relatively straightforward to design, it's strongly encouraged that an experienced Deployment Designer be utilized for even moderately challenging deployments, and especially for complex deployments. If the deployment is estimated to take over two hours in-store, for more than 100 stores, and on a national level, it is a good investment to bring in an experienced Deployment Designer that can ensure every aspect of the deployment is covered and designed appropriately.

When planning the attendee list for the workshop, include all the people that will be expected to participate in the implementation of the Deployment Design. Without them, critical details necessary to make the deployment successful will be missed. In retail technology deployments this includes every individual in a Project Management role, all Technical Architects and Engineers involved in the technology solution being deployed (it may also be necessary to include technology vendor representatives), Store Operations, Logistics personnel, a Help Desk Supervisor and specific Representatives that will assist in implementing the final Deployment Design, Project Sponsor, and any other key decision makers. If possible, involve lead Deployment Technicians. If using a third-party provider for the deployment, include all the appropriate people from that organization, as well.

Hire a Technical Writer or designate someone on the team to take detailed notes during the entire workshop. There will be a tremendous amount of information collected and decisions made, so ensure everything is captured properly. During weeklong workshops, it is helpful to have two people documenting, ensuring that nothing is missed.

Set up the workshop so that all participants can be in the same room together. This meeting is crucial to the success of the deployment, so having people in person is preferable. These tend to be all day events, and much can be missed by remote participants. Spend the money to bring in experts during this critical time.

The complexity of the deployment determines how long the workshop will take to complete. At a minimum, schedule two days for relatively straightforward deployments and up to a week for highly complex deployments. If more time is needed, take it. Time spent designing is time saved during the deployment.

Prior to the workshop, have all the material to be covered organized and placed in a form that is easily consumable for the attendees. The atmosphere for the workshop should be interactive and very collaborative. Make sure everyone involved is introduced and understands each person's role in the project. Keep people engaged throughout the workshop by using props like fun Post-It Notes, colored pens and markers, white boards, and other items that can be hung up around the room during the discussion.

Also, give breaks throughout the day, allowing people to interact on a more personal basis and to brainstorm together. As the Project Workshop Leader, it is important to manage the participants and keep everyone on topic. If there are areas where there are outstanding questions or decisions that need to be made, create a place to write down action items and topics that need to be returned to later. Put this list in a place where everyone will be reminded of what is outstanding.

For the workshop agenda, include everything related to the *Who*, *What*, *When*, *Where*, and *How* of the project— essentially, all the components of this book in detail. There is nothing too insignificant to consider and no detail that does not need to be captured.

The output of the Design Workshop is the Deployment Design and should be shared with all workshop participants. Everyone should review the design and provide feedback to the Project Leader. It may be necessary to have follow up meetings to finalize some specific areas, but the design should be 90% complete by the end of the workshop. Make sure to set due dates when the design is sent out for review, ensuring everyone remains on schedule. Once all feedback is received, obtain final sign-off from the Project Sponsors and ensure final approval is obtained from all required decision makers.

Once final approval is received, implement the Deployment Design. Ensure that any changes made to the Deployment Design after the deployments start are properly captured, tracked, and implemented. This allows the exact design details that were most effective to be properly captured and used for future deployments.

Key Takeaways:

- *Utilize a two- to five-day Design Workshop that includes all Team Members responsible for the success of the project, including experts and decisions makers, to discuss and document every aspect of the deployment.*

- *The output should be properly captured by a Technical Writer or documentation expert, as it will be the direct input to the Deployment Design.*

- *The time spent during this design phase will pay for itself many times over during execution.*

CHAPTER 4 – SCOPE DEFINITION

Learning Opportunities:

- *Why a complete Scope Definition is critical to the Deployment Design.*

- *The components that make up the Scope Definition.*

- *Methods to confirm the Scope Definition.*

Although the need to define scope clearly may seem obvious, it is surprising how many Retailers do not take the time to define everything initially. At best, Retailers often define which stores are included, when the deployments must be completed, a very general overview of exactly what is being deployed, who will run the project, and how many Deployment Technicians are needed—maybe. These same Retailers are often surprised when the deployment fails in the early stages or they experience uncontrolled cost overruns.

The Scope Definition is the starting point for designing the deployment. It is imperative to document fully what is to be included and to outline clearly the activities that are to be performed, where exactly they are to be performed, who is responsible for completing each activity, any purchases related to them, when the work will be done, when the store is available for the work to be done, any special equipment that is required to complete the work, and any assumptions that are to be applied to the activities.

Developing a well-defined scope means aligning expectations with actions and building controls that provide fiscal accountability. In other words, clearly documenting everything that is going to be done allows for the control of costs for the deployment by establishing very clear boundaries. To do anything differently only sets the deployment itself up for failure and overrunning the budget—all because the scope was not clearly defined up front.

It can be just as important to document what is not included in the scope. These are usually referred to as assumptions and they can be critical when it comes to evaluating out-of-scope work during the deployment. By clearly documenting assumptions it is known to all that the work was evaluated and determined that it would not be included. Work may be specifically omitted from scope due to timing, financing, or resource availability.

The exercise of Scope Definition is relatively straightforward. Simply define the *Who, What, When, Where,* and *How* of the deployment to be included in the Deployment Design. Depending on the depth, breadth, and complexity of the specific deployment, this can be a daunting task. To get started, the following outlines key considerations for each of the components.

The Who

The *Who* defines all the resources required for the deployment project. Most projects will involve more resources than just Deployment Technicians and a Project Manager. Consider other resources that could be crucial to the success of the deployment: Logistics Coordinators, Help Desk Technicians, Technical Subject Matter Experts, Master Schedulers, Financial Analysts, Procurement & Configuration Specialists, Store Operations, Executive Oversight, and a Change Control Board.

Additionally, for every resource required to execute the project, the expectations of each role and how many of each resource needed must be defined. Specify whether he or she should be an internal or external resource, permanent or contract, and part-time or full-time on the project.

Chapter 5 covers how to define *The Who* for the deployment.

The What

The *What* aspect defines the logistics required for the deployment. Outline clearly what specific equipment will be deployed to the store along with how the equipment will be acquired and shipped to and from the store. If configuration is required, this must be outlined specifically, along with how long it is estimated to complete the configuration. State if asset information needs to be collected and provided by Procurement or Deployment Technicians.

Along with stating the specific equipment to be deployed, be specific about how it will be deployed. Specifying clearly how long it is believed the deployment will take is part of the Scope Definition. If not clearly defined, this can lead to out-of-scope work during the deployment which can add up very quickly if scoped and controlled improperly.

Chapter 7 covers how to define *The What* properly.

The When

The *When* is the first part required for scheduling. Make sure to define clearly the earliest date the deployment can begin and the latest date to be completed. This information drives scheduling and plays a key role in defining resource requirements for the entire deployment.

To ensure minimal in-store interruption, state clearly when Deployment Technicians can complete their scope of work and the process for them to follow if they run into issues that could lengthen the deployment. Also, define blackout periods when no deployments can take place in the store.

If this will be applicable to the deployment, define a ramp up and ramp down period. To ensure support resource utilization is as predictable as possible, define the minimum and maximum stores that can be deployed per day.

Chapter 6 covers *The When* component of scheduling.

The Where

The *Where* is the second component of scheduling. Include the exact store list with store numbers, phone numbers, and full addresses that are included in the deployment. This allows for a coverage analysis to be conducted, facilitating proper planning of resources and predictive scheduling.

It is also very helpful to provide the type of footprint for each store (assuming there is some form of standardization across the store portfolio) and current as-built floor plans, all of which allow the Deployment Technicians to know where infrastructure already exists, and they can optimize their time in the store.

Specify any special considerations for a store so these can be built into the Deployment Design accordingly. If the deployment itself has specific requirements (e.g. a dock, specific size storage area, etc.), document how stores that do not meet the requirement will be handled.

Chapter 6 covers *The Where* portion of scheduling.

The How

The first part of *How* outlines how the deployment itself will be executed. Provide as much information as possible on the steps to be taken as part of the deployment (it is best to include the actual deployment script). Indicate if a pilot or proof-of-concept is to be completed as part of the scope and how many stores should be included.

The second part of *How* defines the support model to be used. If it is a relatively small deployment, perhaps everything can be handled by a Project Manager. For larger and more complex deployments, it may be necessary to establish a Command Center and a Help Desk. Specify how these need to be setup and run.

The last part of the *How*, and usually the most overlooked, is to identify the governance model to be used, reporting requirements, escalation points, the decision-making process, and how a Change Control Board should be utilized.

Chapters 8 and *9* outline all the considerations for defining *The How*.

Scope Confirmation

The best way to confirm scope and to understand fully all the costs involved is to complete a pilot or proof-of-concept phase for a few, carefully-selected stores. This should be done very early in the design process to help confirm the scope. Be sure to include the following components in the pilot:

- Attendees should include a Project Manager, Technical Subject Matter Experts (SMEs), Deployment Technicians, Time Keeper, Technical Writer, and Team Leads.

- The scope of the pilot itself should include: timeline, specific stores, types of deployments, who will execute which stores, how equipment will be delivered, how tasks will be timed, and how the issues encountered, and results, will be documented and resolved.

- Try to include one of each type of deployment, if possible, and a reasonable percentage of the total number of stores in the deployment.

- While stores selected are, usually, within proximity to the IT Team to save on travel costs, it can pay to travel to geographically diverse stores if they can prove to be helpful in vetting out the overall deployment scope.

- With so many people in attendance, the deployment of pilot stores may have to be done before or after store hours to minimize disruption. However, it is very important for all parties to attend to ensure success.

- It is very important to track, to time, and to document everything in detail, and to take pictures along the way.

This information will be used to solidify the scope, to finalize contracts if a third-party provider will be used, to improve deployment scripts, and to optimize workflows.

- Before starting, define a kill point. If the deployments are fraught with issues, (i.e. taking much longer than estimated or are not successful) the remainder of the pilot should be cancelled. At that point, it would become necessary to go back to the drawing board.

- Don't take any action for granted. The deployment should be the end-to-end process including clean up and check-out.

- After the pilot period, all details captured, and lessons learned are incorporated into the Deployment Design.

It is obvious that these key considerations generate a cascading set of questions, which should then be used to complete the Scope Definition fully. Once the Scope Definition is developed completely, it is fed directly into the Deployment Design. Therefore, the Deployment Design will only be as good as the Scope Definition. Make sure it receives the attention it deserves.

Key Takeaways:

- *The Deployment Design starts with a Scope Definition.*

- *A well-defined scope aligns expectations with actions and builds controls that provide fiscal accountability.*

- *A pilot or proof-of-concept is an effective way to confirm the Scope Definition.*

CHAPTER 5 – RESOURCES

Learning Opportunities:

- *The Resource Types required for most deployment projects.*

- *The responsibilities for each Resource Type.*

- *How to determine quantities needed for each Resource Type.*

The number of resources necessary in even a seemingly small deployment can be staggering, but without the right people, in the right place, at the right time, the cost will be multiplied. Having the right number of resources for a deployment is important but having the right type of resources can be even more important. Most of these resources are the face and the voice of the project. They interact on the phone and in person with the store personnel; it is very important that they provide superior customer service.

So, what are all the resource types that need to be considered for any deployment project? While each project has unique characteristics, they will use the same set of foundational tasks and resources. Here are the key resource types, along with the important aspects of each role.

- **Executive Oversight / Sponsors / Steering Committee**

  o Key decision makers.

  o Represents the overall vision.

  o Provides approvals for material changes and associated risks.

  o Takes decision-making burdens from the Project Manager and Deployment Team.

  o Provides budget and associated cost approvals.

  o Provides overall strategic guidance.

  o Assists with navigation of approvals.

- **Change Control Board**

  o Reviews all requested changes with material impact and assesses risks.

  o Approves or rejects requested changes.

  o Ensures proper implementation of approved changes.

  o Meets on a regular cadence but can call emergency meetings if needed.

- **Program / Project Management**

  o Responsible for the overall success of the deployment.

- o Regardless of size, every deployment requires at least one Project Manager.

- o For very large deployments that span several years, more than one PM may be required and potentially a Project Director.

- o Should have specific experience with retail technology deployments and the ability to manage large groups of resources that are geographically diverse.

- o The lead Project Manager should be involved in all aspects of the project to ensure consistency and cohesiveness.

- **Project Coordination**

- o Serves as a communication channel between the various Deployment Team Members and the Project Manager to ensure all issues are addressed.

- o Performs coordination activities like call aheads, check-in and check-outs, and issue escalations.

- o Can be used to supplement PMs, especially for data collection, follow ups, and reporting.

- **Logistics Coordination**

  o Responsible for ensuring that all resources and equipment are where they should be, at the time they should be, and in the correct condition.

  o Troubleshoot logistical issues.

  o Practice proactiveness as much as possible.

  o Build in checkpoints to support proactiveness.

  o Must be organized, a logical thinker, and not afraid to talk on the phone.

- **Scheduling Master**

  o Responsible for all aspects of scheduling deployments and all associated milestones.

  o Assigns the appropriate Deployment Technician based on service areas and location proximity criteria.

  o Handles reschedules and revisits based on pre-established Service Level Agreements (SLAs).

  o Escalates missed schedule milestones proactively and participates in resolution to keep the deployment on track and to avoid a reschedule scenario.

- **Help Desk**

  o Provides various levels of technical support to the Deployment Technicians during the deployment.

  o Expected to stay organized, to handle potentially chaotic environments, and to think under pressure.

  o May also handle call aheads, equipment tracking, and follow up customer satisfaction surveys with stores.

  o Must have experience with customer service, not afraid to talk on the phone, troubleshooting skills, ability to follow scripts, and project-specific technical skills.

- **Deployment Technicians**

  o Serve as the face of the deployment to the store.

  o Responsible for completing all aspects of the deployment in the store.

  o May work as part of a Deployment Team or alone.

  o Must be able to follow instructions, communicate in a professional manner, and troubleshoot technical issues.

  o In most deployments, travel and before-and-after-hours availability are required.

- **Staging / Configuration / Kitting**

  o Prepares equipment to be deployed by picking, packing, configuring, and kitting for shipment.

  o Must be able to follow a configuration script and assemble hardware components.

  o Must be organized and detailed.

- **Repair / Disposition**

  o Handles recovered equipment from stores to either be repaired for redeployment or sold.

  o Properly disposes of unneeded equipment and provides appropriate disposition documentation as required.

- **Order Processing / Equipment Management**

  o Responsible for ensuring that equipment orders are placed properly so that equipment arrives at the time it is needed.

  o Ensures equipment is received and handled properly prior to deployment.

  o Resolves any issues related to equipment availability.

  o Implements the appropriate asset management requirements.

- **Quality Control**

  - Ensures the quality standards established for the deployment are implemented and followed properly.

  - Provides guidance on quality standards to all Team Members.

  - Documents and resolves any quality issues discovered.

- **Documentation Specialist / Technical Writer / Librarian**

  - Ensures proper documentation of all aspects of the deployment.

  - Establishes and implements version control either manually or via a version control system.

  - Responsible for establishing and maintaining a repository or library and implementing change control processes for all documents.

- **Technical Practitioners / Subject Matter Experts**

  - Provides technical expertise and guidance for deploying the designed technology solution.

  - Provides solutions for stores or scenarios that do not fit the standard deployment.

- o May serve as Level 4 or Level 5 when a technical issue cannot be handled by the Help Desk.

- o Heavy utilization usually required during Deployment Design and during the early phases of the deployment but diminishes as more deployments are completed.

- **Store Operations**

  - o A crucial participant in the Deployment Design.

  - o Provides guidance on how stores are set up and operated.

  - o Facilitates communication to the stores related to the deployment, associated impact, and timeline.

  - o Handles issues with store personnel.

- **Administration / Contracts / Financial**

  - o There is usually some level of administrative support personnel necessary to support the deployment including, Human Resources, Billing, Budget & Contract Administration, and Legal.

  - o Usually utilizes existing staff of an organization.

Once the resource types have been determined, define key requirements for each one. This information will be used to vet potential candidates, ensuring that only the resources matching the requirements are assigned to the deployment. At a minimum, the following should be defined clearly for each resource type:

- Job Description
- Years of Experience Required
- Specific Skills Required
- Education and Certification Requirements
- Employment Requirement (W2, 1099, or Vendor)
- Dedicated or Shared Resource (i.e. Full-Time or Part-Time)
- Specific Tools Required
- Location of Resource (Local or Remote, Onshore or Offshore)
- Special Security Requirements or Clearance
- Compliance Certification Requirements (e.g. HIPAA Training, OSHA Certified, etc.)
- After-Hours Availability Requirements
- Medical Requirements (e.g. Flu shot required to work in hospitals)
- Specific Project Training Required

The final, and most difficult, aspect to determine is the quantity required of each resource type. While this may seem straightforward, it is not always the case with every resource type. Deployment Technicians are usually the most challenging to determine because the quantity is based upon the number of deployments and the timeframe in which they need to be completed.

For example, let's say a deployment project that covers 5,000 stores needs to be completed in a 24-month period. That is approximately 209 stores per month (5,000 stores / 24 months = 208.33 stores per month). Now let's say the install takes two days per store, and work can only be done in the store Monday through Thursday of each week. Assuming there are four weeks in each month with four working days in each, there are eight install windows per month (4 weeks X 2 install windows per week = 8 install windows per month). With 209 stores per month over 8 install windows, approximately 27 stores per install window must be deployed (209 stores per month / 8 install windows per month = 26.125 stores per install window). *Figure 5-1* provides a template to use for these calculations.

	A	B	C
1	# Stores	5,000	User Input
2	# Months	24	User Input
3	Total Stores per Month	208.33	B1 / B2
4	**Total Stores per Month**	209.00	**Rounded**
5	# Days per Install	2	User Input
6	# Days Available a Week	4	User Input
7	# Install Windows per Week	2	B6 / B5
9	# Weeks per Month	4	User Input
10	# Install Windows per Month	8	B7 * B9
11	**# Stores per Install Window**	26.12	**B4 / B10**
12	**Total Stores per Install Window**	27.00	**Rounded**

Figure 5-1

If the install requires two Deployment Technicians each, at least 54 Technicians will be needed per install window for 24 months. Keep in mind this scenario does not account for blackout periods, a ramp up period inside the 24-month deployment timeframe, or extra resources needed in the event of illness, attrition, travel issues, etc. The trick is to have enough Deployment Technicians to complete all the scheduled deployments but not too many they must be placed on the bench because there isn't enough work for them to do.

It can get even more complicated if the deployment requires Deployment Technicians with different skill sets and there are different deployment store footprints that require different skill sets. The bottom line is to make sure the quantities are based on sound assumptions for the specific deployment and to vet the assumptions with not only the Project Management Team but with the Technical Team, as well.

In terms of managing project resources throughout the deployment, create a continuous improvement feedback loop for all Team Members, so they can learn from each other and incorporate a process to handle gaps when issues arise temporarily. If there are Team Members who are detrimental to the success of the deployment, remove him or her immediately. These difficult situations don't improve and can have a negative impact on a project leader's credibility.

Defining resource requirements clearly is a critical component of the Deployment Design. Spend the time necessary to ensure the vast array of people needed to complete a successful retail technology deployment are properly assembled.

Key Takeaways:

- *Having the right types and number of resources defined in the Deployment Design is a critical component.*

- *Within the Deployment Design ensure the resource requirements are built on solid assumptions and can be scaled as required.*

- *Include in the Deployment Design the guidelines and processes to properly monitor and manage the resources effectively.*

CHAPTER 6 – LOGISTICS

Learning Opportunities:

- *The components of Logistics requirements.*

- *The various aspects of Logistics requirements.*

- *How the Logistics requirements are incorporated into the Deployment Design.*

Logistics can mean different things to different people at different times. For retail technology deployments, it involves all the movement and activities that take place during the deployment lifecycle. This includes all the equipment, packing material, transportation, tools, configuration, disposition, check-in and check-out, asset tracking, entry and exit procedures, and sign-off.

Logistics requirements are an important aspect of the Deployment Design and must include all the activities expected to be performed. There are two main logistics requirement categories to consider: asset management and tracking and in-store activities and processes.

Asset Management & Tracking

This category covers everything related to any equipment required for the deployment.

- **Configuration & Staging** - If equipment for the deployment needs to be configured prior to deployment, the requirements must outline how and when that configuration will be completed and who is responsible for configuring. Three ways to handle configuration include:

  1. Configuration completed and confirmed before shipment to the store.
  2. Configuration completed and confirmed by a third-party in their facility prior to shipment to the store.
  3. Configuration completed and confirmed by the Deployment Technician upon arrival at the store (i.e. the equipment will be configured in the store).

  Regardless of who is responsible for configuring and confirming the equipment, the configuration instructions should be documented and tested thoroughly. It is best to test and confirm the equipment configuration as early in the Deployment Design process as possible.

  When providing the details of the configuration, include an estimated length of time to complete the configuration, along with any requirements for configuration (e.g. software media, Internet connectivity, etc.) and how the lack of one of these requirements may impact the configuration process.

The staging of equipment can cover the entire lifecycle of the deployment, from receipt of the new equipment from the manufacturer, asset tracking, configuration, storage, order processing, pick, pack, ship, and generation of the proper documentation.

- **Shipping** - Defining the shipping requirements for the equipment includes the list of specific devices and components to be shipped to the store, the associated weight of each piece of equipment to be shipped (if unknown, this can be determined with the make and model of the equipment to be shipped), any special considerations for shipping (e.g. 8' network cabinet cannot be shipped via FedEx or UPS, so a transportation company will need to be utilized), preferred carrier or transportation company, how soon prior to deployment the equipment should be shipped (i.e. how long can the equipment sit in the store before it needs to be deployed), and who will be required to cover the cost of shipping.

  Define the requirements for shipping of legacy equipment removed from the store post deployment. The requirements to be considered are the same as the new equipment being shipped. Additionally, be specific as to the party responsible to prepare, deliver, report, and confirm the legacy equipment has been returned properly.

If equipment needs to be transported from one store to another by a Deployment Technician, outline specifically how this should be handled and if there are special considerations for this transportation. This method of equipment movement is fraught with issues and risks so careful consideration should be given before using.

- **Packaging** – Similar to shipping requirements, it is important to list the requirements for how the packaging is to be handled and who is responsible to provide and complete the packaging process. For new equipment, the packaging is often handled by the manufacturer, but if there are several pieces of equipment, it may be decided to place them on a pallet and shrink wrap them to ensure all the equipment arrives together at the store. If this is the case, designate who is responsible for providing these supplies, arranging the equipment on the pallet, and securing for pickup.

  For legacy equipment being returned post deployment, specify how the legacy equipment should be packaged for return. If boxes and packaging material cannot be used from the new equipment, designate how this will be provided in-store and who is responsible for covering the cost.

- **Transportation** - There are various types of transportation that can be used depending on what type and how much equipment needs to be transported. These include traditional package carriers, such as UPS and FedEx, as well as transportation companies like Pegasus and CalArk. Transportation could include truck rentals, such as Penske, Budget, or U-Haul. These rentals can be used to go to stores to retrieve legacy equipment.

  Requirements should include the specifics on what transportation mode should be utilized for the deployment, any specific carriers or transportation providers to be used, who is responsible for arranging and providing transportation, and who is responsible to cover the cost of transportation.

  Liability considerations must be given to scenarios where Deployment Technicians will be requested to drive a rental truck to either deliver equipment or to retrieve legacy equipment. It is important to conduct the proper amount of due diligence to mitigate any risks this approach generates.

- **Equipment Handling** - Once the equipment arrives at the store, specific instructions on handling must be given to the Store Manager. This includes designating where the equipment should be stored, how to confirm all equipment arrived, who should have access to the equipment, and how the equipment is to be turned over to the Deployment Technician upon arrival.

  Once the Deployment Technician arrives at the store, specify how he or she will obtain access to the equipment, how the equipment should be accounted for (e.g. record serial number for asset tracking, etc.), how to handle missing or Dead-On-Arrival (DOA) equipment, and how the equipment should be removed from its storage location.

- **Disposal** - Equipment disposition can be handled in several ways, including: disposal, destruction, resell, repaired and redeployed, or donated. The type of disposition to be utilized for each piece of legacy equipment should be outlined specifically.

  When considering the type of disposition to be used keep environmental, security, and financial aspects of that disposition in mind. Donation of legacy equipment is a popular option for most organizations, but exposure of sensitive data must be expunged completely to ensure no risk of data breach to the donating organization.

- **Chain of Custody** - this refers to the chronological documentation, or paper trail, showing the seizure, custody, control, transfer, analysis, and disposition of physical or electronic evidence. Essentially, chain of custody is the unbroken trail of accountability for the electronic device from its removal to its final destruction.

  Often, chain of custody is required for any device that contains sensitive data, like personnel records, customer credit card information, or personal health information, that, if found, could cause a financial loss to the affected party. Confirm what legal requirements are necessary for chain of custody documentation and ensure that all requirements are communicated and satisfied.

- **Destruction of Equipment** - There are scenarios where equipment will need to be destroyed to ensure that no sensitive data can be reclaimed. Usually, this pertains to hard drives in computers, and servers. These devices should be crushed, melted, or destroyed permanently, ensuring it can never be used again.

If equipment is required to be destroyed, a certificate of destruction, or death certificate, must be provided. This will include a description of the equipment, serial number, means of destruction, destruction date, and the organization that completed the destruction.

- **Compliance Requirements & Documentation** - All documentation required for equipment acquisition, staging, configuration, packaging, shipping, tracking, receiving, installation, and disposal must be defined clearly to ensure that all data elements required for that documentation are in place from the initiation of the deployment.

    In addition, any compliance requirements, either legal, organizational, governmental, or industrial, must be outlined specifically and communicated.

In-Store Activities & Processes

This category covers everything related to the activities that occur in the store as part of the deployment.

- **Entry and Exit** - It is common with invasive work that the Deployment Technician complete the deployment before or after store hours. In this scenario, designate who will grant access for the Deployment Technician into the store and what his or her expected arrival time will be.

Also, there should be an escalation procedure in place if the Store Manager does not arrive at the designated time or does not open a locked door for the Deployment Technician. Time spent waiting on access to the store may be considered out-of-scope and incur additional costs.

- **Check-In and Check-Out** - Always require that Deployment Technicians check in with a Project Coordinator or Help Desk upon arrival at the store. Their onsite time should be monitored closely to ensure that the deployment does not fall behind schedule.

  Depending on the length of the deployment window or time constraints, it may be required that the Deployment Technicians check in at specific milestones to ensure they are where they should be and to address any issues promptly.

  Deployment Technicians must also check out with a Project Coordinator or Help Desk. There should be an opportunity to confirm that all work was completed, all deliverables have been collected and that there are no lingering issues once they've departed the store.

- **Turn Up and Test Out** - In some deployments, a turn up process is required to be followed which utilizes both the Deployment Technician as well as another member of the Deployment Team. Usually, this is because the Technician does not have visibility or access to the parts of the system that are required to bring systems online. In this scenario, have specific instructions for the Deployment Technician to call and have the turn up completed. Outline how long a Deployment Technician should wait for assistance for turn up and the escalation process to be followed if this time threshold is exceeded.

  There should be specific instructions for the Deployment Technician to test what has been deployed along with a process to follow if the testing fails. This could include calls to a Help Desk for assistance, engaging a Team Lead or Subject Matter Expert, or following troubleshooting steps. Regardless of the method used to handle failed tests, clear limits must be placed around length of time to troubleshoot before engaging the appropriate escalation procedure.

- **Sign-Off** - It is best practice to require some type of formal sign-off from the store once the Deployment Technician has completed all work. This can include a checklist of items to be confirmed by the Store Manager, but at a minimum should require their signature, along with the date and time of sign-off. If there are issues, a process should be in place for reporting and correcting any issues noted in the deployment.

- **Rules of Engagement** - For all deployments, regardless of size and length, specific rules of engagement should be documented for all Team Members involved. This includes all the specific processes outlined in this chapter, along with the expected behavior for interaction among the various parties.

  In the event issues arise due to a Team Member's interactions, escalation procedures should be established so everyone is aware of how these situations will be handled and who they will be handled by.

Key Takeaways:

- *Logistics requirements include all the activities to be performed from start to finish, including asset management and tracking and in-store activities and processes.*

- *There are a significant number of moving parts for any size deployment so including in the Deployment Design ensures the deployment will not end up with missing and unaccounted for equipment and meet all security, chain of custody, and governance requirements.*

- *By fully designing store activities by the Deployment Technicians, proper expectations can be set with Store Operations in detail and well in advance of the deployment.*

CHAPTER 7 – SCHEDULING

Learning Opportunities:

- *The different methods of Schedule Management.*

- *All the schedule drivers that must be considered when designing the schedule.*

- *How to define and properly establish a Milestone Schedule.*

Imagine a national Retailer with a new project. The entire portfolio of 2,000 stores needs the wireless infrastructure updated quickly, and there are only six months available to complete the project. A site survey and installation review must be completed, equipment configured and shipped and a three-day call-ahead conducted *prior* to the install. Afterwards, the old equipment must be picked up from the store and disposed of properly.

Assuming there are four weeks in every month, there are 24 weeks total (not accounting for holidays or blackout periods) in which to complete the deployment. That's roughly 84 deployments a week. Assuming deployments can be done five days a week, that's 17 deployments a day.

Consider this. If there are seven major milestones that must be completed before, during, and after the deployment, that's 14,000 scheduling milestones that must be tracked throughout the life of the deployment. With so many milestones to develop, manage, and track it is imperative that the schedule be properly designed. Before designing the schedule, there are several considerations to keep in mind including the type of schedule management to be used, schedule drivers for the specific deployment, and available deployment resources.

Schedule Management

Understanding who controls the schedule is very important. For example, does the Retailer control the order and timing of the deployment schedule, does a third-party provider control the schedule, or is it a combination of both? If all the deployment work is to be completed by the Retailer's internal resources, then the Retailer, obviously, will control the schedule. However, if the deployment work is being completed by a third-party provider, then it must be decided if the Retailer will allow them to set and control the schedule or if they will mutually manage. Let's look at each type of schedule management in more detail.

- **The Retailer Controls the Schedule**

    In this scenario, the Retailer will have complete control of the schedule. This means the Retailer will develop, manage, and communicate the schedule. If a third-party provider is completing the deployment work, they may request specific scheduling criteria or changes to the schedule, but the Retailer is not obligated to take the requested criteria into consideration or to make the requested changes. However, the third-party provider is still required to deliver to the specified schedule that the Retailer provides. In this scenario, the Retailer carries the risk of ensuring that schedule changes are controlled tightly, so that additional fees are not incurred from the provider due to last minute changes, poor, or miscommunications.

- **Third-Party Provider Controls the Schedule**

  Here, the third-party provider designs the schedule and has complete control. This form of schedule design and management is very rare as Retailers do not want to give up complete control of the schedule to a third-party provider normally; although, there may be times where this may be appropriate. In this scenario, the third-party provider carries the risk of designing and managing the schedule and will be held to a higher standard of schedule compliance since they not only control the schedule but also the resources to deliver the schedule. Any missed or miscommunicated schedule gaffes will cause poor customer satisfaction, as well as potential fines due to inconvenience.

- **The Retailer Develops a Proposed Schedule and Transfers to Third-Party Provider**

  In this situation, the Retailer designs the proposed schedule, then transfers management to the third-party provider. This transfer of schedule management may be accompanied with specific criteria for making any changes. With this form of collaboration, it is critical that a schedule change management process be established and followed to ensure that both parties understand and agree to the change. Any changes made outside of a formal change management process can cause poor customer satisfaction, scheduling mishaps, and potential fiscal penalties.

Schedule Drivers

Once schedule management control is determined, outline specifically all the elements that drive the schedule. For example, is the schedule design based on the availability of Deployment Technicians, the desires of the stores, or geography? Each schedule driver presents is own associated challenges, but it is important to understand what criteria to use when designing the schedule.

- **Start & End Dates**

  Before the start of any schedule design, the deployment start and end dates required must be determined. This sounds obvious, but much disagreement can arise when establishing these dates. Have a discussion around these dates and understand how they were determined. Most of the time, these dates are driven by external business requirements and have nothing to do with the more reasonable length of time it takes to complete the deployment.

  In a perfect world, the end date would be based on the scope of work, how long it takes to complete each store, how many Deployment Technicians are available to complete the work, where they are located, and how far they must travel. The reality is that the work must usually be completed in time to correspond to a business event. For example, by Black Friday or by a compliance deadline. If the business requirements given are so unrealistic that it is believed that the deployment cannot be successful, negotiate with the business to try and break the deployment into phases. While this can incur more costs down the road, it can prove to meet the business need and allow enough time to deploy the required technology successfully.

- **Store List**

  Before designing the schedule, make sure to obtain a complete list of all stores to be deployed along with the associated physical address. This information will be needed to align the Deployment Technicians with the stores accurately, which then serves as an input to the schedule. A Deployment Technician can only be in one place at one time, so understand the availability in relation to the stores that are within their coverage or service area.

  Additionally, the store list should identify any stores that may require special accommodations or additional expenses due to the remote location. For example, Alaska, Hawaii, and Guam may require flying a Deployment Technician to the store if a local resource cannot be acquired to do the work. Shipping to these stores can also require additional costs and logistics considerations.

- **Days & Times for Deployment Work**

  Specifically define the days and hours during which the deployment work can be completed in the store. For example, most Retailers dictate that work can only be done Monday through Thursday, 8:00 AM to 5:00 PM local time. If it is a 24-hour store, the Retailer may allow for deployments Sunday through Thursday, 9:00 PM to 6:00 AM local time.

Depending on the type of deployment work performed, it may be necessary to complete off hours so as not to impede business. If overhead work is being done or a lift is required to complete the work, a safety hazard to store personnel and customers can arise, so deployment work must be completed when the minimum number of people will be impacted.

Also, consider the Retailer personnel that must be in the store with the Deployment Technician to allow for entry, to facilitate logistics and physical store access, and to provide final sign-off. If work is being done off hours, this may require that store personnel be paid overtime to accommodate the schedule, increasing deployment costs. It is critical to make sure the right people are in the right place at the right time as cost effectively as possible.

- **Blackout Periods**
  Blackout periods are specific times when no deployment work is permitted in the store. These periods should be

defined clearly for all stores. Typically, Retailers don't allow any deployments in-store from Black Friday until the second week of January. Regional or local blackout periods can include a large event that impacts a few stores, but the Retailer does not want traffic potentially impeded during this busy time. Examples include: Marti Gras in New Orleans, the Indianapolis 500 in Indianapolis, and the Kentucky Derby in Louisville.

- **Deployment Rate**

  While the rate at which deployments are completed is constrained by the number of Deployment Technicians available, the rate is also constrained by the number of support resources. These resources include the Help Desk, Logistics Coordinators, Project Coordinators, and Project Managers. Establish not only the minimum number of deployments to complete per day, and still meet the overall timeline, but also the maximum number of deployments that can be completed and supported per day based on the resources available.

  While there should be a constant utilization of resources, there are several rates typically utilized during the life of a deployment.

o **Pilot Period or Proof-of-Concept** – This is the period when the first deployments are executed to vet the deployment process and to confirm time to task estimates that may have been created earlier in the technology solution planning phase. Depending on the scope of the deployment, this usually ranges from one to five deployments per day and lasts one to four weeks.

o **Ramp Up Period** – This is a period when the number of deployments per day, week, or month increases steadily until the maximum number per period is reached. This is an especially useful phase when there are thousands of stores to deploy in a relatively short amount of time. This allows everyone to get into the rhythm of the deployment gradually and not become overwhelmed, which could put the deployment at risk.

A ramp up period is very common, especially in large deployments, as it allows the Team to work out the details of the deployment as it scales up. Also, it exercises all the established support processes and allows for the identification of process gaps that can be evaluated and resolved quickly without impact to a large number of deployments.

A ramp up period can also reduce the number of revisits, out-of-scope costs, and quality issues because the work is only being completed on a small number of stores. Understand the criteria that should be applied during this period. There may be criteria that must be met for a set of deployments before the number of deployments in the next schedule window can be increased.

o **Full Capacity** – This is the maximum number of deployments per period that can be scheduled and lasts until the maximum number can no longer be reached.

o **Ramp Down** – This is the period at the end of the schedule when there is a gradual slowdown. Often, a ramp down period is used to disengage resources from the project gracefully, rather than to a sudden and complete stop, which causes many resources to be transitioned to other projects. For instance, if there are 125 Deployment Technicians the schedule may be designed to reflect the departure of 25

Deployment Technicians a week for the last five weeks of the schedule.

○ **Reduced Schedule** – This is a period that can be used when there is a need to temporarily reduce the schedule load, but deployments must continue to stay on schedule. This can occur when there are many resources that will not be available to either complete deployments or to provide support for the deployments. If used, this is typically less than a two-week period.

Number of Deployment Resources Available

After obtaining the store list with complete addresses, obtain a list of the Deployment Technicians that will be tasked with completing the deployments. Normally, these resources are dispatched from their home or a location close to their home and not from a centralized location. To plan properly, know the starting point of each Deployment Technician to understand the travel time to the stores in their associated service area. This directly constrains the number of deployments that can be performed in any given schedule window by a specific Deployment Technician.

Once evaluated, it may be discovered that there are no local Deployment Technicians within a reasonable driving distance. In these cases, it may be necessary for a Deployment Technician to drive to the area where the store is located the day or night before or even to fly to the location. Again, all this must be taken into consideration when designing the schedule.

Once the scheduling requirements are determined and a preliminary schedule has been designed, there are several aspects of the schedule that must designed and managed including the milestone schedule, reschedules, and revisits.

Milestone Schedule

Milestones are defined by the Project Management Institute (PMI) as "significant points or events in the project's progress that represent accomplishments in the project". As it relates to deployment scheduling, the milestones are used to develop the Milestone Schedule. To do this, evaluate the entire deployment process and identify each milestone that, if missed, will cause the remainder of the deployment schedule to be at risk of completing on time. By tracking these milestones on the schedule, it ensures that a missed event will become apparent immediately and allow time to handle and to keep the remainder of the schedule on track.

Additionally, include milestones to be completed after the deployment work. This ensures that each store is closed out completely in the proper amount of time. *Figure 7-*

1 depicts an example Milestone Schedule for a major technology upgrade project.

Milestone Description	# Days Before or After Deployment
Site Survey	45 Day Before
Receive Site Survey Results	35 Days Before
Conduct Installation Review & Order Equipment	30 Days Before
Configure & Stage Equipment	25 Days Before
Ship Equipment to Store	20 Days Before
Confirm Store Receives Equipment	15 Days Before
Schedule Lock Down Period	14 Days Before
Complete Pre-Install Preparation	5 Days Before
Call Ahead Confirmation	3 Days Before
Deployment	0 Days Before
Pick-Up Equipment for Disposal	1 Day After
Post Deployment Follow-Up	3 Days After
Satisfaction Survey Complete	10 days After
Confirmation of Equipment Disposal	30 Days After

Figure 7-1

Once a store goes into the lock down period, the equipment has shipped, the Deployment Technician has been scheduled, and all the store preparation work has been completed. Any schedule changes within the lock down period can cause major disruption to the store and, usually, results in additional costs to handle the reschedule. The point at which the schedule is locked down should be included as part of the milestone schedule. Any changes after this date will require authorization from a Change Control Board or Steering Committee.

Reschedules & Revisits

Reschedules occur when an issue has been identified that will cause the store or the Deployment Technician to not be ready for the originally planned deployment. These can be same day reschedules, last minute reschedules, and outside lock down period reschedules. The key distinction is that the Deployment Technician has not arrived at the store for the deployment yet (it is considered a revisit if they have already arrived for the deployment). There are many events that can cause a reschedule.

- No Deployment Technician available for the day and time scheduled.
- Prerequisite work was not completed as scheduled (e.g. backboard not hung, cabling not completed, etc.).
- Equipment did not arrive at the store or the equipment arrived damaged and cannot be replaced in time for the scheduled deployment.

- There is an anticipated weather event (e.g. a hurricane) or there was a weather event (e.g. a tornado) preventing access or safe access to the store.

- The Deployment Technician is unable to gain access because store personnel cannot make the scheduled deployment day and time.

- Access to the store will not be allowed due to an unrelated event (e.g. burglary, break in, fire, water leak, strike, riots or civil unrest, etc.).

- In anticipation of a high traffic period, the store does not want anything or anyone more than necessary to be in the store.

It is important that every effort be made to avoid reschedules. Depending on the causes and frequency, this could have a detrimental effect on the overall schedule if not managed carefully and thoughtfully.

The type of reschedule event will determine how quickly the deployment needs to be worked back into the schedule. Here are some helpful guidelines to consider.

- Outside of the lock down period allows flexibility in rescheduling, especially if the store has not had any type of schedule notification or call ahead.

- If equipment is in the store or the deployment is partially complete, it should be rescheduled within three days.

- If equipment is in transit, determine if it should be diverted to another store, if possible, returned to its origination, or delivered to the store as scheduled. Depending on which option is selected, determines the time horizon in which the reschedule needs to happen.

For very large projects, leave slots in the schedule for reschedules and floating Deployment Technicians to cover these events.

Ultimately, the schedule is what drives the deployment. Make sure every aspect to be considered is understood and everything that must be tracked against the schedule. Do everything possible to make all Team Members adhere to the established schedule. Schedule changes should be the exception; not the rule.

Key Takeaways:

- *To effectively design the Deployment Schedule, the type of Schedule Management, Schedule Drivers, and Deployment Rate must be determined.*

- *Utilizing a Milestone Schedule provides the level of control required to significantly reduce or eliminate the number of reschedules and revisits.*

- *By considering all the different elements that go into designing a Deployment Schedule, schedule changes become an exception rather than the rule.*

CHAPTER 8 – EXECUTION

Learning Opportunities:

- *How to use Site Surveys and Installation Reviews to assess and monitor store readiness.*

- *How to effectively use Call Aheads and Checkpoints to control the in-store deployment.*

- *All the various in-store activities and how they should be managed and controlled.*

This is where the proverbial rubber meets the road. Execution includes all the activities that must be completed to ensure that the deployment is completed flawlessly. The following outlines the various activities to be considered as part of overall execution.

Site Surveys

One of the best ways to reduce risk for retail technology deployments is to make sure that all the various aspects of every store in the portfolio are known. There are few things worse than sending a pallet of equipment to the store to be deployed only to discover that there is no dock available or that space is unavailable to secure the equipment until the Deployment Technician arrives. Another risk could include realizing that the workstations in the store that need to be upgraded don't have the correct version of the operating system or the correct configuration. By using site surveys effectively, all this information will be known in advance, avoiding a reschedule, needless interruptions to the store, additional costs, and frustration by all involved.

The purpose of a site survey is to confirm the physical aspects for the planned deployment, to collect store information for the purposes of future use, and to identify any special requirements or issues that need to be resolved prior to the deployment. This can include physical aspects of the store, an inventory of equipment in the store, checking the operating system and configuration levels of currently installed devices, specific location of devices, and even the collection of device serial numbers.

Performing a site survey on every single store may be unnecessary, so it is important to outline the criteria that must be met for a site survey to be conducted. It will need to be determined if the site survey can be conducted over the phone with store personnel or require a visit by a Deployment Technician to walk through the store and communicate with the store personnel. It's also possible that a combination of both may be the best option. For example, a preliminary phone survey may be conducted and, based on the results, determine if an in-store site survey needs to be conducted.

It is best to have a written script to follow during the site survey, but at a minimum, provide a checklist. The script or checklist should include every aspect of the store that needs to be checked, confirmed, and documented along with pictures that are required to be taken. Additionally, it will need to be determined if the results will be recorded manually, then communicated later, or recorded electronically. The latter allows for immediate access to the results.

For planning purposes, determine the length of time the survey is expected to take to complete. Determine if each survey must be scheduled in advance with the store or if the survey can be completed by a specific date, as convenient for the Deployment Technician. Be sure to let the Deployment Technician know what tools he or she will need to bring to conduct the survey (e.g. script or checklist, digital camera, tablet, etc.).

Installation Reviews

The results of the site survey feed directly to the installation review, which is conducted by a set panel of Project Team Members with decision making powers. Depending on the complexity of the deployment, it may not be necessary to conduct an installation review for every store. Be specific about the criteria to be applied to the results to determine if an installation review is required.

Have a set time each week in which to complete installation reviews, and make sure they are completed prior to the schedule lock down period. If for some reason the store needs to be rescheduled due to issues discovered or issues that cannot be resolved in time for the scheduled deployment during the installation review, rescheduling before the lock down period, will avoid any negative impact on the overall deployment schedule.

For stores that require an installation review, have a process in place to review with the Project Team. Each issue identified should be documented thoroughly, along with the resolution and a list of tasks that must be completed prior to the scheduled deployment. There should be a process in place to collect the issues, to communicate to all Team Members, and to assign prerequisite tasks specifically. Additionally, a process must be established to ensure that all tasks are completed prior to the deployment.

The most important aspects of the installation review include ensuring that the deployment does not have to be rescheduled and that it can be completed the first time successfully. Knowledge of the store environment prior to the deployment is one of the keys to successful retail technology deployments.

Call Aheads

Everyone knows that communication is important when executing projects, but this is especially true for retail technology deployments. One of the key communication tools used is the call ahead. While it seems simple enough, a call ahead must be designed correctly at specific points within the overall deployment timeline.

Call aheads can be used for many purposes, but the most common use is to confirm store readiness, including equipment arrival, scheduling of resources, all pre-work completed, and confirming that the Store Manager can arrive at the scheduled time to open the store. There are four aspects to call aheads that should be considered for any retail technology deployment.

1. Timing

When designing the deployment, include the specific points in the timeline to perform the various call aheads. For stores, this is usually a week prior to the equipment arrival and then again two days before the scheduled deployment date. If, for some reason, the store is rescheduled or requires a revisit to complete the deployment, reset the call ahead status so they can be called again at the appropriate times.

A word of caution—do not design too many call aheads into the timeline. Balance the right number of calls to communicate the information needed for a successful deployment, while not overburdening the store. Calling the store constantly to confirm information or to remind employees of information will not be effective and can lead to confusion. Working closely with Store Operations can ensure that the touch points are frequent enough for success.

Call aheads should not just be used with stores but with the Deployment Technicians, as well. The more aggressive the schedule, the more important these calls can be. There is nothing much worse than the Store Manager showing up early to let a Deployment Technician into the store only to have the Technician late or absent. To avoid this, a Deployment Technician should be called at least two days in advance of their scheduled deployments. In the event he or she cannot meet the schedule, there will be time to find a replacement without having to reschedule the deployment.

2. Consistency

To be consistent and to ensure clear communications, a script should be developed that all Team Members executing call aheads will use to convey the necessary information and to confirm requirements. Scripts should include exactly what should be said and the specific list of items that must be confirmed during the call ahead. When training personnel to conduct these calls, stress the importance of not improvising. While there is no desire for the caller to sound disengaged in the process, it is important to make sure that each Store Manager hears the same information. Store personnel talks to each other and shares information among their sister stores, so inconsistent information can cause unneeded misunderstandings. If there is a need to confirm something unique that does not apply to every store, include that in the script. Let that store's personnel know that they are one of a few stores to have this unique aspect and that this needs to be confirmed with them. It is strongly encouraged to develop these scripts with Store Operations to ensure there is alignment with the proper communication style to which the stores are accustomed.

3. **Tracking**

   It can't be stressed enough: track *everything* related to a deployment, including the information discussed and the feedback received during the call ahead. This can be a manual form to be completed during the call ahead or an application in which to enter the information but create a system for everyone to follow. Suppose a deployment covers 1,000 stores, and each store requires four call aheads each. That is 4,000 calls, assuming call backs to talk to the correct person are not necessary. Don't leave the information to someone's memory or "shorthand". Everything needs to be documented in a fashion that is clear and concise to anyone reviewing the call ahead.

4. **Processes**

Based on the complexity of the deployment, and the various types of store footprints, it may be necessary to have different scripts for different stores. The workflow and associated criteria must be designed to determine which call ahead script and checklist should be used for which type of store. A process should be in place to handle issues that are identified during the call ahead.

For example, what happens if the Store Manager says she cannot have the deployment in her store on the scheduled day because of inventory or because the equipment has not arrived at the store even though the tracking information says it has arrived? These situations can cause a great deal of stress to handle, or even fall through the cracks altogether if there is not a predefined process to follow. There must be a process for these issues to be escalated to avoid any disruption to the deployment schedule.

Checkpoints

Checkpoints are used to track the Deployment Technician's progress. This ensures that any delays in progress are identified quickly and escalated. Checkpoints should be used for all deployments, regardless of size.

Specify ahead of time at what point in the deployment these checkpoints should occur. If the deployment is short in length (less than four hours) it may be possible to require a check-in and a check-out checkpoint only. A longer deployment may require checkpoint times, which are determined by the specifics of the deployment.

For example, if it is estimated to take four hours to install half of the point-of-sale devices, have the Deployment Technician call in at the four-hour mark. If they should have completed six yet they've only completed two, there is a problem. By performing this checkpoint, there is time to do something to get the deployment back on track. In this scenario, another Deployment Technician may be sent to the store to help.

A Help Desk or assigned Project Coordinator should be available for the Deployment Technician to call upon arrival, at pre-established checkpoints, and upon departure. He or she should call if running late or behind on their deployment.

All information should be recorded by the Project Coordinator and tracked. The information tracked should be used to identify trends related to deployment timing, Deployment Technician reliability, recurring issues (either technically or with a specific Deployment Technician), and issues with the store. This information should be used for billing and auditing purposes.

If a checkpoint does not occur when expected, the Project Coordinator must track down the Deployment Technician to determine if there is a problem. Alarms must be put in place to ensure the Deployment Technician is checking in appropriately. Otherwise, it will never be known if the deployment is at risk.

In-Store Guidelines

General guidelines should be in place for Deployment Technician behavior, along with Retailer-specific guidelines, if applicable. There should be a zero-tolerance policy for not following these guidelines.

Specific attire and appearance for a retail deployment includes: collared shirt (always tucked-in), jeans or khakis with no holes or rips, safety shoes or dark colored sneakers, possibly a hard hat in new construction, union required attire, and a clean and neat appearance (no body odor). Smoking onsite by Deployment Technicians should follow the Retailer's guidelines to their employees.

A point of contact should always be designated for the store. This is usually the Store Manager. He or she is responsible for approving the Deployment Technician's activities while in-store. While in-store, the Deployment Technician should be sensitive to impeding customer traffic. He or she should be aware of how ladders, lifts, and equipment can block access for customers. However, the Deployment Technician should block off aisles when using a lift during store hours.

When moving ladders and lifts around the store, the Deployment Technician should pay attention to surroundings and customers (if work is performed during store hours). Lifts should be lowered completely before moving and help should be enlisted if moving lifts or ladders is required.

Deployment Technicians should keep areas secure at all times. There may be occasions when handling a safe key and work in locked offices is necessary. They must be vigilant about maintaining the key and the associated security of where they are working. No one should be allowed access that is not pre-approved.

While in the store, the Deployment Technician may be asked for specific information or to complete additional tasks by store personnel. He or she should never commit to anything without gaining approval from the Project Manager. It is critical that out-of-scope work not be performed without prior approval.

Deployment Checklists & Guides

Deployment checklists, guides, and scripts should always be used. Test instructions continuously until they can be completed with no issues before the deployment starts. If the deployment changes, the instructions must be updated and retested. The instructions should be tested by various people with various levels of technical abilities and understanding.

Format and mark all documents as confidential. Deployment Technicians should be instructed never to leave documents unattended. In addition to the deployment instructions, be sure to include other important information, such as:

- How to complete check-in and check-out.
- How to contact the Help Desk, Project Coordinators, and Project Managers.
- How to escalate out-of-scope events.
- How to handle missing equipment.

- How to dispose of trash.
- How to interact with store personnel.

Always use a fully tested deployment script—no matter the size of the deployment. When developing a script, retest the script until it can be fully executed without issue.

Usually, the original script is written by a technical person or Technical Writer. Once written, the Project Manager or Project Coordinator should execute the script and update with their findings. They should repeat the process until they feel it is solid. Then, have someone who is not familiar with the script execute and update with their results. If they must ask even one question about the process, it will have to be repeated.

Continue this process until the person executing the script does not have to ask any questions and completes the deployment correctly. Obviously, this can be a very tedious process if it is a lengthy deployment (e.g. full day or multi-day deployment), but the payback will be realized whether there are 20 or 200 deployments being executed on the same day. There is nothing that will stop a deployment faster than an inundated Help Desk because Deployment Technicians cannot complete the script.

Checklists can be used to supplement the detailed instructions and for deployment completeness. They can also be presented for sign-off to the store personnel.

There may be multiple deployment documents for different deployment configurations. Owners should be designated for each one so that there is an identified expert that knows everything about the document. They are also responsible for making updates when required throughout the deployment project.

Have an established process for document version control. The version numbers should be printed clearly on each page of the document and only the current approved version should be available to the Deployment Technician. Establish the process for publishing new versions during the deployment. This should be incorporated into their deployment checklist for confirmation of the correct version of the deployment instructions.

Test out and turn up procedures should be used when necessary to complete and confirm the deployment. This may be done via written instructions, over the phone, or a combination of both. There may be a different group from the Help Desk that performs this process so that Deployment Technicians are not waiting in-store to complete the deployment. There needs to be a well-established process in place to handle problems that are encountered during this process.

Out-Of-Scope Work

Out-of-scope (OOS) work is any work that is not included in the work order, deployment instructions, or deployment checklist. Usually, OOS work involves material or more than 15 minutes to complete. It may impact the deployment in such a way that is causes a reschedule, a revisit, or a complete cancellation.

One of the most common OOS events is Deployment Technician wait time. Typically, if a Deployment Technician is waiting more than 15 minutes, an OOS event has started. Common causes of wait time include:

- Lack of access to store or to specific secured areas of the store.
- Call back from the Help Desk or a Project Coordinator to assist with an issue encountered.
- Assistance logging into a device by store personnel.
- Equipment not in the store.
- Waiting on a second Deployment Technician to arrive.

A specific process should be developed to handle OOS events. Often, this outlines how the Deployment Technician is to contact the Project Coordinator or Project Manager, so he or she can review cost, time, impact, and risks. The OOS work is then approved or denied, and that is communicated back to the store and to the Deployment Technician.

Usually, Deployment Technicians are not approved to complete OOS work without gaining the prior approval of the Project Coordinator or Project Manager. Typically, store personnel are not empowered to make these types of decisions.

When completing OOS work, documentation is critical. The following data points must be fully captured:

- Detailed description of requested work.
- Purpose of the work.
- Cost of the work.
- Time estimated to complete the work.
- Risks, if any, and associated mitigation strategy.
- Deployment impact.
- When work is to be completed.
- Who approved the request.

Communication must be complete, but also streamlined so the deployment is not impacted negatively.

Clean-Up & Exit Procedures

During the process of documenting the deployment process, list everything that is contained within all boxes and everything that is used to ship the equipment. Every item that comes in for the deployment, and is not deployed, must be accounted for and disposed of in some manner. Old equipment should have explicit instructions on how to be handled.

No trash should be left for the store personnel to deal with. It must be determined if trash can be disposed of in-store or if it must be taken off site and disposed of at another location. Recycling may be required and designed as part of the deployment.

Sign-off by store personnel should be required as part of deployment completeness. If there is an issue with sign-off, an escalation procedure must be established to release the Deployment Technician. Deployment Technicians must confirm that any remaining issues are captured properly and communicated to the Project Team. These issues should be reviewed by the store personnel providing sign-off to ensure agreement on the final deployment state.

Document fully all chain of custody requirements, as well as the handling of devices with sensitive data. All pictures required should be listed specifically. Photos can be taken with a phone or a digital camera and must be provided to the

Project Team within a specific time period after the deployment. Any missing pictures or documentation may require a revisit at the Deployment Technician's cost. All documents and pictures should be secured and treated as confidential information.

Key Takeaways:

- *Store readiness activities are a critical component of a successful Deployment Design.*

- *Designing the appropriate checkpoints can ensure the execution of work in-store is tightly controlled and monitored.*

- *Not impeding customer traffic, minimizing impact to store operations, and providing superior customer service should be stressed throughout the Deployment Design.*

- *Prepare for issues and have predefined processes to handle them expeditiously.*

CHAPTER 9 – SUPPORT

Learning Opportunity:

- *The various methods that can be used to provide deployment support.*

- *The importance of establishing escalation procedures.*

- *How Continuous Process Improvement is vital to the overall success of the Deployment Design.*

Every deployment project needs support. Support can include the establishment of the following: A Deployment Help Desk, a Command Center, strategically deploying Team Leads, and utilizing Subject Matter Experts.

Deployment Help Desk

A project leader has designed and planned a complex retail technology deployment down to the last detail. The equipment configurations and deployment script have been perfected. Call aheads have been performed and the Deployment Technicians have been trained. The deployment schedule has been worked and reworked and sign-off has been received from Store Operations. A communication plan has been developed and daily status reports to communicate progress have been established.

Finally, the day has come for the first group of deployments. All the sudden, the project leader starts getting calls from the Deployment Technicians with questions and realizes quickly that support cannot quickly scale to support all of them throughout the deployment. This is because the support that is required with any sized retail technology deployment was not properly designed. While there are multiple aspects of deployment support, including the establishment of a Command Center, deploying Team Leads or SWAT Teams strategically, and utilizing Subject Matter Experts, one of the main components of a successful deployment support plan is to utilize a Deployment Help Desk. This ensures there is a fully staffed and flexible support group in place to help the Deployment Technicians and even store personnel, if needed. There are different aspects to consider when designing a successful Deployment Help Desk.

- **Internal or Outsourced** – The first decision to make is whether the Deployment Help Desk will be handled internally by existing Help Desk Specialists within an organization, by adding temporary Specialists, or outsourcing to a third-party. If there isn't already an internal Help Desk, it can be daunting to set up at least a temporary Deployment Help Desk, so outsourcing may be the best solution. An existing outsourced provider may be able to provide these services, or a new provider can be hired for these specific services.

- **Location** – Will the Deployment Help Desk be at the retail corporate headquarters, at a third-party provider's location, offshore, onshore, or near-shore? The location will determine how information will be communicated. Communication is key, so consider carefully how information will be shared.

- **Call & Ticketing Systems** – Is there an existing Help Desk call and ticketing system that can be utilized for the Deployment Help Desk? If so, ensure the system can handle the anticipated increase in call volume. Make sure the existing ticketing system can handle the types of calls that will be received during the deployment. Are the proper categories and workflows available to utilize? If not, can they be set up, and how much effort will be required?

- **Hours and Coverage** – When establishing hours of operation, consider the various time zones in which Deployment Technicians will work and the earliest and latest that calls may be received. Consider other tasks that may be handled by the Deployment Help Desk (e.g. call aheads, equipment delivery tracking, and follow up tasks) and during which hours these tasks need to be performed. This is often driven by the store's hours of operation but consider when people may call in for assistance. This helps to determine the resources required to staff and scale properly.

- **Support Levels** – Normally, there are levels of assistance available on a Deployment Help Desk (e.g. Level 1, Level 2, Level 3, etc.). The higher the level, the more complex and technical the issue. Define these levels, what types of issues are expected to be handled by each level, and how the handoff will happen between levels. Determine how unsupportable issues at any level will be escalated and handled.

- **Expertise Requirements** – After establishing the support levels required for the deployment, determine what education, training, certification, and experience requirements are needed. The best way to do this is to create a job description for each type of resource. Be specific, and plan to apply this criterion when selecting Deployment Help Desk Specialists.

- **Resource Requirements** – Next, determine how many of each resource type will be required to staff the Deployment Help Desk adequately. This is not an exact science, especially initially, but if valid assumptions are applied and the ability to scale up or down as needed is built in, it will be successful. Consider if the resources will be dedicated to the deployment project fully or shared, along with anticipated churn rate over time. It may take several iterations and adjustments to get this right, so make sure this is built into the design.

- **Service Level Agreements** – Service Level Agreements (SLAs) are used to establish expectations around response times for specific types of events to control costs and to keep the Deployment Schedule in line. If outsourcing the Deployment Help Desk, be sure all SLAs are documented specifically in the contract in detail, along with the process for SLA measurement, frequency of measurement, and financial penalty for not meeting the SLAs. If using an internal Help Desk, do the same thing, foregoing the financial penalty. SLAs can be unique to the Deployment, but at a minimum, consider the following:

  - o Average Time to Answer
  - o Average Hold Time
  - o Average Time to Call Back
  - o Average Time to Resolution

      ○   Average Time to Escalation

- **Reporting** – Specify the types and frequency of reporting. These reports can be critical for monitoring call volume and ensuring that issues are handled promptly and correctly. Look for trends by correlating specific issues or call volumes to specific times of the day, specific areas of the country, and specific Deployment Technicians. Being specific from the start, guarantees the correct data points are being collected all along the way.

- **Escalation Processes** – Make no mistake about it—there will be unplanned events. Prepare for these situations by creating a robust and succinct process for handling those situations where no operating procedure was developed previously. Avoid the Deployment Help Desk Specialists spending too much time spinning their wheels and make sure that they know how to handle the unplanned.

The plan for the Deployment Help Desk is a critical part of the Deployment Design. Include the Team Members that support Store Operations currently, as they bring a great deal of experience and knowledge to the table and can be vital to the design. If the considerations above are given the appropriate time and thoughtfulness, the results when the deployment takes off at full speed will be pleasing.

Command Center

A Command Center, or War Room, is a centralized model of deployment support and provides a significant benefit for very large and complex deployment projects. It includes Team Members who represent all areas of the deployment and work in the same location (and preferably the same room). This facilitates real-time collaboration and becomes the hub of all preparations and deployment activities. This makes mobilizing very easy when major issues occur.

Usually, this approach is implemented for very large deployments with very short timeframes for deployment. All members of the Command Center are expected to be onsite and dedicated full time to the deployment through its entire life. As the deployment ramps down, members of the Command Center should be released once all their responsibilities have been fulfilled, documented, and approved by the Project Manager.

Team Leads

Team Leads are responsible for leading and mentoring other Deployment Technicians. They should understand all aspects of the deployment intimately and be able to move between stores effectively. Team Leads can be used when the assigned Deployment Technician runs into an issue in which he or she needs assistance. This could be due to an unexpected issue or running behind on the deployment. The additional resource can help get the Deployment Technician back on schedule.

Team Leads should be the most experienced and savvy of the Deployment Technicians. They may be utilized to fill-in for other Technicians on an emergency basis. This requires that they be very flexible and willing to travel with minimal notice.

Team Leads can be crucial in keeping the deployment schedule on track and, even, salvaging deployments that might otherwise need to be rescheduled.

Troubleshooting & Escalation Procedures

To help Deployment Technicians manage their time in-store effectively and stay on schedule, layout specific guidelines for them to follow. Ambiguity can lead to additional costs because of the additional time Deployment Technicians spend waiting.

One of the most important procedures to establish is wait time criteria. This dictates how long a Deployment Technician can wait for something before escalating for assistance. Examples of wait time include: access to the store, check-in, equipment arrival, Deployment Help Desk call back, and test out.

If there are found to be emerging trends that cause wait time across multiple deployments, complete a root cause analysis. By understanding what causes delays, processes can be put in place to negate any future wait time.

Wait time events should be closely monitored. If the Deployment Technician's work is impeded by store personnel, escalate immediately and handle sensitively. This type of issue is handled most often by the Project Manager rather than the Deployment Help Desk or even called into the Command Center for resolution. For each type of wait time scenario, document all possible use cases and the workflows to be followed for each. Have enough escalation points so that the Deployment Technician can reach someone within five minutes.

When a Deployment Technician encounters an issue with the deployment and goes into troubleshooting mode, there should be an associated time threshold. Once the threshold is exceeded, he or she should be required to escalate and either request additional troubleshooting time or assistance from the Deployment Help Desk or a Team Lead. If additional troubleshooting time is approved, set a new threshold for the additional time. Continue with these checkpoints until another course of action is needed to keep the deployment on schedule.

Make sure that all work during this troubleshooting period is documented fully in the event there are questions later or additional costs for this time. Capture the details of the work, who requested the additional time, who approved it, and how much time was spent.

Post Deployment

Once the deployment is complete, gather every artifact that must be collected and documented. In the deployment instructions include a checklist of all documents that must be collected, including:

- Deployment Checklists
- Signed Off Deployments
- Pictures
- Shipping Documents
- Chain of Custody Documents
- Birth and Death Certificates

When setting up a documentation and library of artifacts, it is preferable to have a web-based application, so that it can be accessed readily by all Project Team Members. SharePoint is a popular choice for these types of repositories, but keep in mind that VPN access may be required by Deployment Technicians. Other choices include BOX, Google Docs, Microsoft Azure, or a custom web site.

The contents of the library should be organized so that Team Members can locate needed documents easily and know where to place their deployment artifacts. Every document should be catalogued and given a unique document ID.

All artifacts must be submitted to the Project Team within a specific timeframe. The Project Coordinator is responsible for ensuring that all artifacts are collected, catalogued, and filed. Every effort should be made to obtain any missing artifacts.

Remaining issues after the deployment is complete should be reviewed by the Deployment Help Desk. They should make sure each issue is assigned appropriately and an anticipated response time is established. They must continue to track until all issues are resolved.

Information Systems

Most Retailers, and even third-party service providers, do not have a system to track every aspect of the deployment project. Most of the time, large numbers of Excel spreadsheets and Word documents are passed back and forth. Inevitably, this causes confusion, and someone makes a critical decision based on the wrong version of a document. While anyone can certainly be very successful executing deployment projects using these types of documents and a SharePoint site, organization and detail are vital to ensure that there is a way in which to manage changes effectively.

In the best scenario, information systems are in place to collect and report the information in an organized and detailed fashion. This system should be available to all Team Members and access controlled at the user level. Not only does this information benefit the current deployment project, but the information collected can be used to facilitate future technology deployments.

Governance

Governance is a crucial aspect of any deployment, so it shouldn't be ignored or glossed over. It is used to ensure that the deployment stays in control, both from a timing perspective, as well as from a fiscal perspective. Governance provides visibility to the sponsors and executives who are responsible for the project, along with oversight for all Team Members, ensuring the overall vision is met.

Establish a Steering Committee. Regardless of the size or complexity, the Steering Committee will provide oversight and the decision-making services required throughout the life of the deployment. The members should represent all departments and areas impacted by the deployment. Meetings should be held at least once a month, or as needed if issues arise that require immediate attention.

A written status report should be provided to all members, and minutes should be recorded and distributed within 48 hours. Any negative impacts to deployment costs, timeline, or resources should be reviewed and approved, assuming the Change Control Board has approved. Usually, the meeting is led by the lead Project Manager.

Another very helpful tool is to establish a Change Control Board (CCB). While CCBs have been used traditionally with software development projects, they serve a very important purpose for any type of technology project. The CCB is responsible for handling all the changes to the deployment that have the potential for a material impact, such as on cost or the overall timeline.

An existing CCB can be used, if appropriate, or a project-specific CCB can be established. Typically, this group meets once a week on a regular cadence but can meet more frequently or on an emergency basis, if needed. The CCB is responsible for establishing the basic elements of the change process, including:

- What is under change control and what is excluded.
- How changes are requested.

- Who has the authority to approve or reject changes.
- How decisions, upon approval or rejection, are documented and disseminated.
- How changes are implemented, and the implementation recorded.

No change request should be outstanding for more than a week, and each request should include the following elements:

- The nature of the change.
- The reason for the change.
- An assessment of the urgency of the change.
- The expected impact of the change.
- How the change will be implemented.

Continuous Improvement Process

The Continuous Improvement Process (CIP) is an ongoing effort to improve processes, usually on an incremental basis. This process allows for the improvement of processes continually, rather than never adjusting once implemented. As new processes are implemented, it will be realized that adjustments must be made to work with real-life events rather than based on theory exclusively.

Establish owners for each process, document, and artifact and make them responsible for keeping track of all observations, suggestions, and testing results. Ensure every document is version controlled to allow for improvements to be implemented properly.

Changes should be reviewed and approved by the Continuous Improvement Process Team and implemented and communicated via an established process.

Within a deployment, it is critical that changes are communicated properly to the Deployment Technicians in the field. It is very helpful early in the deployments to schedule regularly recurring calls to review improvements with the Team Leads and Deployment Technicians, helping to ensure understanding and to gain additional feedback.

The cycle for improvement review is driven often by the size and scope of the deployment, but regular reviews should be established.

Reporting Metrics

It is very important that every aspect of the deployment be tracked—even if it is believed that the data point is not required. Inevitably the one data point not collected will be needed.

Reporting requirements need to be defined very early on in the deployment. Identify the various reports that will be needed and the consumers of those reports. Understanding these requirements early ensures that data elements needed for reporting properly are being collected and tracked.

Establish a master system of record that contains the single source of truth. This can either be manual or electronic, but it should be identified and communicated to the Project Team. Implement processes that ensure all data has been entered in the appropriate timeframe into the system of record.

Another important aspect of the data elements is ensuring accuracy and integrity. One rule to use is "if it doesn't make sense, it's most likely not true". Question everything and overlook nothing.

Key Takeaways:

- *Designing the support is as important as designing the deployment execution itself.*

- *Utilizing a Deployment Help Desk, Command Center, Team Leads, and fully documenting escalation processes and procedures will ensure the deployment is properly supported.*

- *Ensure governance is implemented from the start to monitor project control and maintain fiscal accountability.*

CHAPTER 10 – DEPLOYMENT DESIGN

Learning Opportunities:

- The key aspects of creating the Deployment Design.

- How to collect and properly implement changes to the Deployment Design.

- How to implement the Deployment Design once complete and approved.

The Deployment Design Workshop has been conducted, all the open issues and questions have been closed out; now, there is a mountain of documentation to organize. It's time to pull all the information together into the final Deployment Design. The Deployment Design details the workflow and the specific steps that will be used to deliver the deployment. This design may be presented as a workflow document, as a written document, or as a combination of both, but it should provide all the information that every member of the Deployment Team will need to execute a successful deployment.

Here are some key aspects to consider when creating the Design Deployment:

- The document should be arranged in categories that make sense for the type of deployment being conducted. The same types of categories outlined in this book can be used or different categories can be made. Just make sure that everything is easy to read and follow.

- The use of swim lanes or RACI (Responsible, Accountable, Consulted, Informed) Charts can be very beneficial, especially when there are numerous hand-offs between different groups during a workflow.

- Be consistent with the terms used throughout the Deployment Design. For example, if the resources that handle the logistics are referred to as 'Logistics Coordinators' make sure to always refer to them as such. Don't refer to the resources as 'Logistics Managers' or just 'Logistics' later in the document.

- People are the most important aspect of the Deployment Design, so it can be very helpful to include a roles and responsibilities matrix or an organizational chart and to include the specific names of the resources assigned for clarity.

- One important element that many people overlook when designing a deployment is to include the assumptions used for the Deployment Design. Inevitably, someone will ask why something wasn't included or why certain elements are designed the way they are—the assumptions will explain why. This is important when working with a third-party provider on the deployment. All assumptions can be placed in one section or align them to the various topics they apply to throughout the Deployment Design. Here are some examples of the types of assumptions that might be a part of the Deployment Design:

o The third-party provider has complete control of which stores are scheduled if they are inside the project start and end dates as directed by the Retailer.

o No Technician training will be required.

o Equipment will be confirmed prior to Deployment Technician arrival.

o Retailer will provide tracking information for shipments to ensure a reschedule fee is not incurred due to delayed equipment.

o If equipment is not at the store upon Deployment Technician arrival, the store deployment will be rescheduled at the earliest time available.

o The Technician time to task in-store does not include:

  ▪ Confirmation of equipment serial numbers.

  ▪ Looking for old equipment in the store.

  ▪ Testing of new equipment.

  ▪ Repackaging of old equipment.

- All detailed deployment and test documentation will be provided, and version controlled.
- Documentation will be provided five days prior to its required use.
- Wait periods or troubleshooting of more than 15 minutes will be considered out-of-scope.
- Return Merchandise Authorization (RMA) instructions will be provided if any equipment is found to be defective out-of-box.
- Retailer will provide all packaging material required to return old equipment.
- Retailer will provide pre-paid return label to return old equipment.
- There is a FedEx drop-off location within 10 miles of the scheduled store.
- Stores cancelled with less than 48 hours' notice will be charged a reschedule fee.
- There will be three holidays when deployments will not occur.
- Technician training will be provided remotely and will only require one hour to complete.
- Sign-off sheets will be provided electronically within five (5) business days of completed deployment.

- No additional deliverables will be provided beyond the sign-off sheet.
- Implement document version control, and make sure every page of the document indicates the version number.
- Mark as confidential and ensure that it is secured properly, as it can contain very sensitive information that must be controlled at all times. It should only be shared with the appropriate people at the appropriate time. Old versions should be shredded when no longer needed.
- Include a Change Log (at the end of the document) and maintain throughout the Deployment Design's life. The Change Log should track all of changes to the document once the baseline is approved by the Project Sponsor or Executive. Each change should include the following elements:

  - Date of the change.
  - Who requested the change.
  - Who made the change.
  - Description of the change.

- Each page should be numbered and have a confidentiality reference.

- If using logos in the Deployment Design document, have permission to do so, and, if so, follow all the branding guidelines required (including trademark references).

Once the Deployment Design is completed, it should be presented to the Project Sponsors and Executives for final approval and acceptance. While the Deployment Design will be continually updated throughout the life of the deployment, it is critical to have a baseline Deployment Design approved. This aligns expectations properly and places the deployment down the path of continuous improvement.

After approval is received and the baseline is set, the Deployment Design should be turned over to the Project Management Team for execution. They are responsible for the proper implementation and ongoing maintenance of the design as a part of their overall Deployment Plan.

Based on the size of the Deployment Design to be implemented, a Mobilization Workshop may need to be conducted. During this workshop, the Project Management Team will assemble all the resources, as specified in the Deployment Design and then implement all the various aspects of the design itself. The length of time required for the Mobilization Workshop depends upon the level of effort required to implement the Deployment Design.

At the completion of the workshop, deployments should be ready to start. This is where all the hard work pays off. The great news is that, since the process of creating the Deployment Design was used, this is really the smoothest part of the deployment project. It should run like a very well-oiled machine, and any issues that do arise can be handled expeditiously with little or no impact to the overall deployment. Now, everyone can really enjoy the fruits of all their labors!

Key Takeaways:

- *The Deployment Design includes all the elements required to successfully execute the deployment project.*

- *No detail is too small, and every aspect should be considered in detail.*

- *Fully document assumptions and decisions to support the Deployment Design as well as the Continuous Process Improvement program.*

- *A Mobilization Workshop can be used to implement the Deployment Design for execution.*

GLOSSARY

As-Built Floor Plans – The floor plans of a building that reflect how the building is built and where specific infrastructure for mechanical, electrical, plumbing, and technology are located. This is useful when completing deployment work that involves any part of the store infrastructure.

Asset Tracking – A method to track the physical location and status of an asset within an organization. Most organizations use some form of software to keep track of the asset description, serial number, location, and current status for financial purposes. Any time there is a change to the asset, the Asset Tracking system must be updated.

Bench – This is a term used to mean that a billable resource within an organization does not have any or not enough billable tasks to work on to keep them fully utilized. Often resources are placed on the bench when moving between project assignments, but they should not remain on the bench long-term (typically less than 30 days).

Birth Certificate – When a device is procured it may go through a process where it is configured, tested, and asset tagged prior to being deployed in a store. At the end of this process a birth certificate is issued to formally document when it was configured, who it was configured by, who it was tested by, and other relevant information. That information is maintained for audit purposes in the event there is an issue with the device.

Blackout Periods – Any time period when no deployment work can be completed in a store usually in anticipation of high volume traffic. Typical blackout periods for Retailers are any holidays (Easter, Independence Day, Black Friday through January), local or regional events (Mardi Gras in New Orleans,

Indy 500 in Indianapolis), and certain store activities (inventory, corporate visit).

Call Ahead – This activity is used to contact a store and Deployment Technician to confirm deployment readiness. For example, a call ahead can be made to the store a few days prior to equipment arrival to make sure they have awareness and have cleared out a place to store the equipment. A call ahead made to the Deployment Technician can be used to ensure they have the correct date, time, and store on their schedule and that they have the right equipment and documentation to complete the deployment.

Chain of Custody – This is the uninterrupted control of a piece of equipment that must be secured until it is destroyed. This usually applies to equipment or devices that contain sensitive data. For example, computers that are being replaced in a store will contain credit card data and other sensitive information. To complete the chain of custody, the hard drive is removed, placed in tamper-proof packaging with the corresponding documentation, handed to a shipper that provides a tracking number, and shipped to its final destination for destruction. This process ensures that the sensitive data is not accessed prior to its destruction.

Change Control Board – A Change Control Board, or CCB, is comprised of a group of decisions makers that provide governance and control over a group of people and processes. Any material changes that are requested are reviewed by the CCB to assess impact and risk to production operations and then either approved or rejected. Only changes approved by the CCB can be implemented into production.

Change Log – A written log that documents all changes made to a document. It usually includes date of change, who requested the change, a description of the change, when the change was made, and who made the change. It also specifically outlines which portions of the document were updated with the change.

Check-In – When a Deployment Technician arrives at a store to begin their work, they should call the Project Coordinator or Help Desk to let them know they've arrived. This ensures that there are no missed arrivals as scheduled and starts the timer for project specific checkpoints.

Check-Out – When a Deployment Technician has completed all their work at a store, they should call the Project Coordinator or Help Desk to let them know all work is complete. This allows for the review of any open issues and ensure they have been closed out properly, confirm that all deliverables have been collected and transmitted, and to officially close out the deployment.

Checkpoints – Checkpoints are utilized in lengthy deployments to ensure that the Deployment Technician is making the right amount of progress and is not falling behind. These checkpoints are predetermined related to the deployment specifics. For example, if it is estimated that it should take the Deployment Technician six hours to install six point-of-sale devices, a checkpoint at three hours should reflect that they have completed the installation of three devices. If they've only completed one after three hours, then action may need to be taken to get them caught up.

Command Center – A Command Center is a physical location where a Project Team works together to manage or provide oversight of ongoing deployments. This collaborative environment is especially useful for complex deployments that span multiple days and projects that span multiple years.

It allows for real-time response to crisis and the rapid formation of mitigation strategies.

Configuration – This is the specific set up of a device to performed as desired. Typically, a device is not configured when it is shipped from the manufacturer. This set up can be completed manually or in an automated fashion prior to sending the device to the store to be installed.

Continuous Process Improvement – Continuous Process Improvement, or CPI, is the ongoing review and updates of processes used for execution. The concept is that as more information is collected and experience is gained through repetition, the better the processes can be made to complete work efficiently.

Coverage Analysis – This analysis is completed to determine where Deployment Technicians are in relation to the stores that are in the project scope. Done early, this analysis will highlight areas that are lacking coverage and identify where recruiting efforts should be specifically focused.

Coverage Schedule – When a Help Desk or other Team works more than a standard eight- or nine-hour day, it requires resources to work different schedules to ensure there is proper coverage. The coverage schedule shows which resources are working which time periods.

Dead-On-Arrival – When a device is pulled out of the manufacturer's box for the first time and it does not power on it is considered dead-on-arrival, or DOA. There is usually nothing that can be done other than to arrange for replacement of the device and return of the DOA to the manufacturer for credit.

Death Certificate – Once devices are no longer needed they must be destroyed using a specific method, especially if the device contains or contained sensitive data. Upon proper destruction a death certificate is generated that shows the method of destruction used, who destroyed it, when it was destroyed, etc. and provided for audit purposes. Many times, these types of certificates are required to meet compliance requirements.

Deployment – All tasks required to install technology devices at a single store or location. This usually includes procurement, configuration, and shipping of equipment, installation and testing of equipment, and any special set up required.

Deployment Design – Outlines exactly how all tasks included in the Deployment Plan will be executed. The Deployment Design is a written document that covers every aspect of the *Who*, *What*, *When*, *Where*, and *How* of the deployment.

Deployment Designer – These are the resources that take a deployment project scope and create a Deployment Design that is then used by the Project Team to execute against. This person usually has a vast amount of experience with multi-site deployments that encompass thousands of tasks, span multiple years, and having varying levels of complexity.

Deployment Plan – Outlines all the tasks to be completed as part of a deployment and their associated level of effort and start and end dates. It does not include how the tasks will be completed—only what will be completed.

Deployment Rate – This is the rate at which deployment project resources will be fully utilized. This is usually stated as a per day or per month figure.

Deployment Script – This is a document that has step-by-step instructions on how to complete a deployment. It includes all steps from the time the Deployment Technician arrives in the store until the tie he or she are ready to leave. It can take a great deal of work to create and properly complete, but it will eliminate questions form the Deployment Technicians while they are in the store.

Deployment Technician – A skilled resource that is trained to complete all associated tasks for a deployment. They usually possess technical skills required to properly perform the deployment but may be trained specifically for a proprietary deployment.

Design Workshop – A series of meetings conducted over several days to gather all information required to complete a Deployment Design. This Workshop should include all the roles required to define the details and make decisions.

Escalation Process – These are clearly defined and documented workflows that handle specific situations and how they should be properly escalated. Escalation processes are used to closely manage a deployment and to ensure that too much time is not spent on an activity when it is not necessary or puts the deployment at risk.

Footprint – A footprint as it relates to retail deployments, represent the physical characteristics of a store. For example, a Retailer may have one store footprint where all stores have 20 point-of-sale registers, where other stores may only have 16 point-of-sale registers and four self-serve registers. Different footprints usually have a different Deployment Design.

Governance Model – This model is the specific process that will be used to govern the project in its entirety. It will outline the reporting structure for decision makers and include specifics on how quality will be maintained throughout the life of the project. The governance model should be established as part of the Deployment Design.

Help Desk – This is a group of people that provide troubleshooting services as support to an end user community or to Deployment Technicians. They can also be used to track deployment milestones, perform call ahead tasks, and ensure the receipt of all deliverables. A help desk can be established on a temporary basis to support a specific project or on a permanent basis to support an organization.

Installation Review – This uses the results of a site survey to evaluate the readiness of a store to receive a retail deployment. It can also identify any special requirements that should be included as part of the deployment. The objective is to eliminate any unknowns prior to the Deployment Technician arriving in the store to complete the deployment.

Kill Point – This is the point during a pilot or proof-of-concept phase when the progress will be stopped based on specific criteria. For example, if there are 10 stores identified for the pilot phase of a project and the Team is not able to complete the first three consecutive installs, this may be identified as a kill point and the Team has to start over with a different approach or manner of execution.

Kitting – This is the process of packaging specific equipment together for deployment. Not only does the kit include the equipment, but all the associated items required to successfully deploy such as cables and connectors.

Legacy Equipment – This is IT equipment that is being removed from a store and being sent for disposal. It has served its useful life and is usually no longer supported.

Lock Down Period – This is the period in the deployment schedule where changes should not be made. The lock down period usually starts once equipment has been shipped to the store and the Deployment Technician has been scheduled for the deployment. Any changes made inside the lock down period can have a significant impact on the overall schedule and cause issues at the store level.

Logistics – This is the movement of resources and equipment and how they are moved or are moving.

Logistics Coordinators – These are the people that track all the details related to the logistics of the deployment. This includes equipment and Deployment Technicians as well as coordinating the schedule with the various parties.

Milestone Schedule – Once milestones are identified for a deployment, it must be determined when in the project schedule they must be completed. Once this is determined, those milestones are added to the schedule, so they can be properly tracked and reported to keep track of store readiness.

Milestones – These are the activities that must be completed to successfully complete a deployment. Usually each milestone is predicated on the prior milestone being completed.

Mobilization Workshop – Once the Deployment Design is complete and properly approved, it is time for implementation. A Mobilization Workshop is a useful tool to facilitate this implementation. It includes all project roles and is usually led by an experienced Deployment Designer or Retail Project Manager.

Out-Of-Scope – Any work that is not included in the Deployment Design, but either becomes necessary or is requested. This is often referred to as OOS work and usually has an associated cost and requires some type of formal approval before it can be completed.

Pilot – A pilot is a group of stores that are used to confirm a deployment script as part of a Deployment Design. It is often completed very early on in the design phase so that time estimates can be established, and assumptions confirmed.

Pre-Install Preparation – This includes all work that must be completed in a store before the deployment can be completed by the Deployment Technician. This can include electrical, cabling, or structural work that may be required to support the Deployment Design.

Procurement & Configuration Specialists – These are the people that work with vendors to purchase equipment required for the deployment and then apply the appropriate configuration so that it can be successfully deployed.

Project Coordinator – This resource is responsible for keeping track and managing all the details related to a deployment. They usually mange at the store, equipment, and Deployment Technician level. They also complete pre- and post-deployment activities such as call aheads, deliverable collection, and client satisfaction surveys.

Project Manager – This resource is responsible for ensure the implementation and execution of the Deployment Design.

They usually participate in the Deployment Design process and then work to ensure its proper implementation. They continue to track the overall project progress and report to the Project Sponsor and other oversight members on a recurring basis.

Project Sponsor – This resource represents the business owner of the project and is responsible for the overall successful delivery. They provide project oversight and have decision making responsibilities. Usually report into an executive committee as part of a governance model.

Proof-Of-Concept – This is similar to a pilot, but usually starts out with more unknowns. It is most commonly used to vet out technical assumptions and ensure that the planned technical aspects of the deployment will work in the store.

RACI Chart – A matrix that outlines the roles and responsibilities of individuals on a project. It indicates their authority level in the project as Responsible, Accountable, Consulted, or Informed.

Ramp Down Period – This is the period when the number of deployments per day, week, or month steadily decreases until all deployments are complete.

Ramp Periods – Periods of time when the total number of deployments completed are either rising or falling, but not at their maximum levels.

Ramp Up Period – This is a period when the number of deployments per day, week, or month steadily increases until the maximum number of deployments per period is reached.

Reduced Schedule – This is a period when the number of deployments per day, week, or month are significantly reduced for a set period due to resource availability. This is usually a very short duration, two weeks or less.

Reschedules – This happens when a scheduled deployment must be moved to another time or date _and_ the Deployment Technician has not arrived at the store. If it is outside of the lock down period, there is usually minimal or no impact. If it occurs inside the lock down period, there can be a financial and resource availability impact.

Return Merchandise Authorization – Also known as an RMA, is the authorization document that is required to return DOA equipment. This allows the manufacturer to properly track the equipment and issue credit once received.

Revisits – This happens when a scheduled deployment cannot be completed by the Deployment Technician and requires him or her to return later to complete the work. There is always a financial and schedule impact for a revisit because it involves unplanned or unforeseen situations.

Scheduling Master – The resource who is ultimately responsible for designing, maintaining, communicating, and executing against the Deployment Schedule. As a part of their scheduling responsibilities they are also responsible for completing coverage analysis, managing the Milestone Schedule, and ensuring Deployment Technicians are recruited based on geographic needs.

Scope Definition – As part of the Deployment Design, the _who, what, when, where,_ and _how_ are defined to create the scope definition. Any activity not included in the scope definition and the Deployment Design are considered out-of-scope and will require additional resources and costs.

Service Areas – These are geographic areas that are a specific size and include all the Deployment Technicians that reside inside that area. For example, a typical service area may be defined as 100 square miles. All the Deployment Technicians would serve all the stores that are located within that service area. By defining service areas, resource needs can be more accurately forecasted.

Service Level Agreements – Also known as SLAs, these are specific measurements that are used to track performance of specific tasks. If the SLA is not properly maintained there can be a fiscal penalty assessed to compensate the one impacted by the delay. For example, an SLA states that 85% of all calls into a Help Desk will be answered within the first three rings or that 90% of all scheduled deployments will be completed within two business days of the original schedule.

Sign-Off – This is the formal, written approval that assigned work has been completed. For example, once a Deployment Technician completes the prescribed deployment, the Store Manager must sign-off confirming the completion of the scope of work.

Site Survey – This survey is conducted when specific information needs to be gathered to complete an installation review or to ensure that all aspects of the store and equipment are known. There is usually a script or checklist developed and used to complete the site survey and the results are collected.

Staging – This is where equipment is set up and configured prior to being sent to the store for installation by the

Deployment Technician. As a part of staging, equipment is configured, asset information is collected, and packaging for shipment is completed.

Steering Committee – A group of people assembled on a project basis to provide oversight, guidance, and decision making. They are responsible to ensure that the project is reflecting the desires of the business objectives as set out in the project charter.

Store Operations – A group of people, usually at the corporate level of a Retailer, that represent the day-to-day actions at a store level. They handle staffing, organization, operational guidelines, and store oversight to ensure all stores in a portfolio run smoothly and efficiently.

Subject Matter Experts – Also referred to as SMEs, these are people that are experts in a specific technical or business area and provide advisory services to a project. They are usually heavily used during the Deployment Design phase, but then are only used on an exception basis throughout the life of the project.

Support Levels – These represent the various issues that are handled by a specifically skilled person on a Help Desk or in a Technical Team. The higher the support level, the more complex the technical issue. For example, Level 1 can handle tasks such as resetting passwords or troubleshooting basic problems. However, Level 4 would handle errors being received during a server installation in a store.

SWAT Team – This is typically a group of people that are very skilled and experienced with specific tasks. As it relates to technology deployments, they can be used to assist in difficult deployments or deployments where the Deployment Technician needs additional assistance.

Team Leads – These are very experienced and knowledgeable Deployment Technicians that provide leadership to other Deployment Technicians and supplement their coverage when necessary. Team Leads frequently provide direction and project updates to the Deployment Technicians that are assigned to them.

Technical Writer – A resource that is experienced at taking complex technical instructions and documenting them in a formal, easy to understand way. They provide the structure and clarity to technical instructions and guidelines. Technical Writers may also be referred to as a Document Specialist or Librarian.

Test Out – This is the process that a Deployment Technician uses to confirm he or she has properly installed the equipment. It is usually completed once the turn up process has been successfully completed. All tests must be completed and passed before the Deployment Technician can request final sign-off.

Turn Up – This is the process the Deployment Technician completes to bring the installed equipment online and operational. It is sometimes performed remotely by another Technician once he or she has remote access. Once turn up has been completed, they must complete the test out phase.

Made in the USA
Monee, IL
02 February 2020

21171699R00076